SOIL CONSERVATION

ASSESSING THE NATIONAL RESOURCES INVENTORY

Volume 1

Committee on Conservation
Needs and Opportunities

Board on Agriculture

National Research Council

NATIONAL ACADEMY PRESS
Washington, D.C. 1986

National Academy Press 2101 Constitution Avenue, NW Washington, DC 20418

NOTICE: The project that is the subject of this report was approved by the Governing Board of the National Research Council, whose members are drawn from the councils of the National Academy of Sciences, the National Academy of Engineering, and the Institute of Medicine. The members of the committee responsible for the report were chosen for their special competences and with regard for appropriate balance.

This report has been reviewed by a group other than the authors according to procedures approved by a Report Review Committee consisting of members of the National Academy of Sciences, the National Academy of Engineering, and the Institute of Medicine.

The National Research Council was established by the National Academy of Sciences in 1916 to associate the broad community of science and technology with the Academy's purposes of furthering knowledge and of advising the federal government. The Council operates in accordance with general policies determined by the Academy under the authority of its congressional charter of 1863, which establishes the Academy as a private, nonprofit, self-governing membership corporation. The Council has become the principal operating agency of both the National Academy of Sciences and the National Academy of Engineering in the conduct of their services to the government, the public, and the scientific and engineering communities. It is administered jointly by both Academies and the Institute of Medicine. The National Academy of Engineering and the Institute of Medicine were established in 1964 and 1970, respectively, under the charter of the National Academy of Sciences.

This project was supported under Agreements No. 59-3A75-4-57, Soil Conservation Service, and No. 59-32U4-4045, Science and Education, between the U.S. Department of Agriculture and the National Academy of Sciences. Preparation of the publication was supported by funds from the W. K. Kellogg Foundation.

Library of Congress Catalog Card Number 86-60330

International Standard Book Number 0-309-03649-9

Cover photograph courtesy of the U.S. Department of Agriculture, Soil Conservation Service.

Printed in the United States of America

Committee on Conservation Needs and Opportunities

M. GORDON WOLMAN (*Chairman*), The Johns Hopkins University
GEORGE W. BAILEY, U.S. Environmental Protection Agency, Athens, Georgia
SANDRA S. BATIE, Virginia Polytechnic Institute and State University
THOMAS E. FENTON, Iowa State University
WILBUR W. FRYE, University of Kentucky
WILFORD R. GARDNER, University of Arizona
GEORGE W. LANGDALE, U.S. Department of Agriculture, Watkinsville, Georgia
WILLIAM E. LARSON, University of Minnesota
DONALD K. McCOOL, U.S. Department of Agriculture and Washington State University
FRANCIS J. PIERCE, Michigan State University
PAUL E. ROSENBERRY, Iowa State University
LEO M. WALSH, University of Wisconsin

Staff

Charles M. Benbrook, *Project Officer*
Carla Carlson, *Editor*
Kenneth Cook, *Consultant*

Board on Agriculture

WILLIAM L. BROWN (*Chairman*), Pioneer Hi-Bred International, Inc.
JOHN A. PINO (*Vice Chairman*), Inter-American Development Bank
PERRY L. ADKISSON, Texas A&M University
C. EUGENE ALLEN, University of Minnesota
LAWRENCE BOGORAD, Harvard University
ERIC L. ELLWOOD, North Carolina State University
JOSEPH P. FONTENOT, Virginia Polytechnic Institute and State University
RALPH W. F. HARDY, BioTechnica International, Inc., and Cornell University
ROGER L. MITCHELL, University of Missouri
CHARLES C. MUSCOPLAT, Molecular Genetics, Inc.
ELDOR A. PAUL, Michigan State University
VERNON W. RUTTAN, University of Minnesota
JAMES A. TEER, Welder Wildlife Foundation
JAN VAN SCHILFGAARDE, U.S. Department of Agriculture, Fort Collins, Colorado
VIRGINIA WALBOT, Stanford University

CHARLES M. BENBROOK, *Executive Director*

Preface

New information on resource conservation has become available in the last few years that will facilitate the analytical assessment of several important conservation issues. The basic source of much of this new information—the 1982 National Resources Inventory (NRI)—is the subject of this report. The 1982 NRI, a nationwide U.S. Department of Agriculture (USDA) survey of all nonfederal lands, contains data on approximately 22 parameters, including physical characteristics of the land and the effects of agronomic practices on soil erosion. The survey was based on observations of approximately one million sites. (The first NRI was completed in 1977. The second was completed in 1982, following the mandate contained in the Soil and Water Resources Conservation Act of 1977. Throughout this report the term NRI surveys refers to the 1977 and 1982 inventories.)

In early 1984, the USDA's Soil Conservation Service (SCS) asked the National Research Council's Board on Agriculture to facilitate the establishment of discussion between the SCS and natural resource experts by providing analyses and recommendations on high-priority conservation issues. The board was asked to evaluate the potential applications of the 1982 NRI, and more specifically to address the following:

- Identification and classification of erodible and fragile soils;
- Identification, methods of measurement, and effects of ephemeral gully erosion;

- Erosion-productivity models and the on-farm total costs of erosion;
- Need for onsite and offsite soil loss tolerance limits;
- Methods to inventory, monitor, and appraise offsite erosion effects;
- Definitions and methods for inventorying urban built-up land and potential cropland;
- Changes in natural resource use and management since the 1977 NRI; and
- Effects and distribution of erosion-control practices.

As part of this evaluation the board convened a workshop in July 1984 on technical aspects of the statistical design and content of the 1982 NRI and a national convocation in December 1984, "Physical Dimensions of the Erosion Problem." The project, including development of this report, was executed by the Committee on Conservation Needs and Opportunities under the auspices of the Board on Agriculture.

The USDA requested this evaluation project, in part, to ensure that maximum use is made of the 1982 NRI data. The committee foresees a wide variety of potential uses for the data. The range of recommendations in the committee's report also suggests that extensive analysis is required to advance conservation. Some of this new work will come under the category of research, some under policy analysis, and some under program evaluation. All such efforts will contribute additional insights into the effective use of public expenditures in support of conservation.

The companion volume to this report, *Soil Conservation: Assessing the National Resources Inventory, Volume 2*, contains 11 technical papers that were presented at the convocation. The papers and accompanying discussion—all based on data from the 1982 NRI—contribute new information on analytical results and methods, specific applications of conservation planning and practice, and innovative uses of data in resource policy and decision making.

This volume contains the committee's major findings, conclusions, and recommendations on the potential uses of the NRI as well as suggestions for future improvements and complementary activities. (The reader should note that in technical discussions throughout this volume, measures are expressed in English rather than metric units. This usage is consistent with the most commonly used NRI data and with erosion prediction models and equations used in soil management.) Chapter 1 identifies five important applications of NRI data and summarizes major findings based on the 1982 NRI. Chapter 2 presents discussion and recommendations reflecting the committee's conclusion that several aspects of NRI data compilation and dissemination should be improved. It also includes recommendations on the use of remote sensing and related technologies in future NRIs, the expansion of NRI coverage to federal lands, and the inclusion of data of potential value in the evaluation of water quality.

The last three chapters of the report include more detailed discussions on technical aspects of data gathering, analysis, and application. Chapter 3 contains information on the process of erosion and the use of equations to estimate erosion. In Chapter 4 the use of erosion-productivity models and on-farm and off-farm erosion damages are discussed. Chapter 5 includes discussions of the application of conservation practices and the need for an improved land classification system.

The committee has commented and made recommendations on a variety of areas involving research, coordination, and administration relating to the NRI and to activities in the field of conservation and land use. Because work related to improvements in the NRI necessitates involvement of a number of agencies, the committee recommends interagency cooperation and coordination. There is much already, but can efforts be improved? The committee also recognizes that many of its suggestions require money and people; both are in short supply. The Executive Summary calls attention to a few particularly important recommendations. The committee is not, however, in a position to evaluate priorities in detail. But it has called attention to a broad spectrum of issues, because they warrant attention, despite the reality of constraints to some immediate solutions.

Much remains to be learned about the processes of erosion and sedimentation, yet what is now known needs wide application. The committee's goal in issuing this statement is to encourage the widest possible use, consistent with good science, of this important new data set—the 1982 NRI.

M. Gordon Wolman
Chairman

Acknowledgments

The committee wishes to express its appreciation to the many individuals who participated in the July 1984 planning workshop and in the convocation "Physical Dimensions of the Erosion Problem," which was held the following December. Their formal presentations, discussions, and informal exchanges provided the foundation of this report. In undertaking this project, the perspective and insights of Ralph J. McCracken, former deputy chief for Assessment and Planning, Soil Conservation Service (SCS), were invaluable. The cooperation of Gary Nordstrom, director of the Resources Inventory Division of the SCS, also was of great aid to the committee throughout all stages of the project.

In addition, the committee expresses its appreciation to William E. Larson, University of Minnesota at St. Paul, and Robert H. Dowdy, U.S. Department of Agriculture-Agricultural Research Service and University of Minnesota at St. Paul, for quickly and efficiently generating and formatting tables based on data from the 1982 National Resources Inventory.

The committee members wish to express special thanks to Carla Carlson and Kenneth Cook for their hours of writing and editing to fashion this report, and particularly to Charles M. Benbrook for his unwavering encouragement and enthusiasm.

Contents

EXECUTIVE SUMMARY xiii

1. SCOPE AND CONTENT OF THE 1982 NRI 1
   Uses of the 1982 NRI 2
   Information from the 1982 NRI 7

2. IMPROVING FEDERAL RESOURCE
   ASSESSMENT EFFORTS 19
   Compilation and Dissemination of NRI Data 21
   Use of Sensing Technologies in Future NRIs 26
   Expansion of NRI Coverage to Federal Lands 30
   Inclusion of Data Related to Water Quality 31

3. THE MEASURES OF SOIL EROSION 34
   NRI Estimates of Sheet and Rill Erosion: The USLE . . 35
   Wind Erosion Estimates 54
   Erosion by Concentrated Flow: Ephemeral Gullies . . . 59

4. ON-FARM AND OFF-FARM CONSEQUENCES
   OF SOIL EROSION 62
   Effects of Erosion on Production Costs 63
   Erosion-Productivity Models 68
   Soil Erosion and Water Quality 72

5. ASSESSING CONSERVATION PRACTICES AND
   LAND CLASSIFICATION SCHEMES 75
   Conservation Practices 75
   The Land Capability Class System 85
   Alternative Land Classification Schemes 91

REFERENCES 95

APPENDIX 99

INDEX 111

Executive Summary

The wearing away of the surface of the land by water and wind can be a gradual but accelerating process. Erosion removes imperceptibly thin layers of fertile soil, rich in nutrients and organic matter, and reduces the ability of plants to thrive in the soil. A reduction in plant growth results in less protective cover for the soil and less plant residue to enrich it. Further erosion occurs, and the process continues. The danger is that erosion can reduce soil productivity so slowly that the seriousness of the problem might not be recognized until the land is no longer economically suitable for growing crops.

Congress first appropriated funds to study soil erosion in 1928. Research stations were located on the most erodible land in the country. Data were collected on soil characteristics, the effects of erosion on crop yield, and the effects of sediments on water bodies. At the same time drought, depression, and the drama of the Dust Bowl combined to focus public attention on soil conservation. The Soil Conservation Service (SCS) was created in 1935 with the passage of the Soil Conservation Act.

Much has been learned about soil and the process of erosion during the past 50 years. But as farming and land management methods continue to change with the advance of technology and new demands on agricultural production, our knowledge of erosion must be put into the perspective of these newer methods and technologies. Progress requires that new information be collected by more refined techniques. This information must be applied to new problems and new situations.

Analysis of data with more precise and reliable methods will provide an increased scientific and technical understanding of soil erosion, transport, and deposition and their effects on crop productivity and water systems—an understanding that can be the basis for control of the process of erosion and conservation of the natural resources of the United States.

A Basic Conclusion

The 1982 National Resources Inventory (NRI) is the most recent of a series of national resource surveys and inventories performed by the SCS. It includes computerized data that represent a statistical sampling of all nonfederal land in the United States. The 1982 NRI and its immediate predecessor, the 1977 NRI, are valuable sources of data. While there are limitations of data and analytical equations in the 1982 NRI, the Board on Agriculture's Committee on Soil Conservation Needs and Opportunities emphasizes that these limitations do not significantly constrain the validity or accuracy of many important applications of the NRI.

Creative use of the NRI allows conservation analysts and planners to project the degree of erosion control that would be achieved by a wide range of combinations of land use and management practices. It is also valuable in identifying the nature and degree of erosion problems that exist across various production regions.

The value to society of soil conservation programs, research, and data compilation efforts stems from the contributions that these activities make to protecting environmental quality and encouraging better management of national resources within agricultural production systems. A continued NRI series has the potential to contribute greatly toward achievement of this basic goal.

Some Priorities

Given the great potential value of the NRI, the committee believes that considerable effort is warranted to ensure that the NRI be made as strong as possible to best serve its various uses. To this end, the report includes a number of recommendations. While detailed priorities cannot be established, four broad areas are emphasized in the recommendations: the scientific bases of the equations used to predict erosion, the enhanced use of the NRI, the relationships between erosion and productivity and erosion and water quality, and the scope of future NRIs. While each is elaborated in subsequent chapters of the report, the thrusts of these four major areas are briefly stated here.

EXECUTIVE SUMMARY

Improvements in the scientific bases of the estimates of erosion will enhance future policy decisions. The magnitude of erosion by wind and by ephemeral gullies requires special attention. Attention must also be given to the way that factors in the soil loss equations are evaluated in the field. Research, analysis, and evaluation by the scientific community in and out of government are needed. The fruits of such work need to be continually included in the policy process.

Supplemental documentation of the 1977 and 1982 surveys, including details of statistical design and measures of reliability and evaluations of the data, is needed to increase the accessibility and potential use of the NRI. Appropriate studies would involve analyses of the data illustrating the applicability at varying spatial scales as well as evaluation of the sources and magnitude of statistical variability inherent in the sampling and collection of the vast amount of data assembled in the NRI.

Estimates of erosion achieve their greatest value when they can be related to the potential effects of erosion on the productivity of the land and on the potential impact of erosion on water bodies. While models have been developed to evaluate where and how erosion affects the potential productivity of the land, additional work is essential to extend and apply such evaluations. Use of the NRI in estimating potential offsite impacts of erosion requires development and verification of links between erosion, transport, and deposition of sediments and associated pollutants in runoff. Evaluations of potential onsite and offsite effects in different areas are essential to the precise delineation of policy choices in conservation.

To achieve maximum value, a national resources inventory should encompass the country as a whole. The inclusion of appropriate data on federal lands—lands that were not included in earlier NRIs—will provide the broader data base essential in addressing national needs and priorities. New observational techniques and data-handling capabilities may enhance future national surveys.

Analytical Applications to Support Conservation

Program Development

The NRI is being used by U.S. Department of Agriculture (USDA) scientists and others as a basic data source in carrying out many important analyses. The committee has reviewed much of the on-going research using the NRI and recommends that further research using the NRI be applied to

- The geographic distribution of the inherent erosion potential of cropland, measured by the physical factors of the Universal Soil Loss Equation (USLE). This work is essential to characterize and understand the severity of erosion-control problems across geographic regions and to evaluate the range of viable solutions;
- The distribution and effectiveness of conservation practices currently used by farmers and ranchers;
- The development of improved land classification schemes appropriate to specific uses based on criteria and data derived from USLE data in the 1982 NRI;
- The relationship of different land classification schemes and conservation measures to alternative policy options designed to conserve the land.

Erosion Prediction Equations

The 1982 NRI has advanced knowledge of conservation needs and, at the same time, highlighted gaps in data collection and empirically based estimation models such as the USLE and the Wind Erosion Equation (WEE). It has raised new questions and has provided a major new data source to use in searching for answers.

To support more cost-effective conservation investments, there is a need for an improved understanding of erosion and more accurate data on all its basic forms. Knowledge of erosion by wind and the concentrated flow of water is particularly limited.

Wind erosion estimates in the 1982 NRI must be used with considerably more caution than the estimates of sheet and rill erosion based on the USLE. While the wind erosion data can be useful relative indicators of wind erosion hazard, at least in the 10 Great Plains states, estimates of wind erosion losses in many regions are not sufficiently reliable for many uses, such as delineating the degree of erosion hazard as a function of estimated rates of wind erosion. A major commitment, guided by a systematic research strategy, will be necessary to correct methodological and empirical shortcomings in the WEE.

There is a need to use the NRI results in gaining a better understanding of another form of erosion known as ephemeral gully erosion, which occurs within natural drainageways and swales. The NRIs contain no quantitative data on ephemeral gully erosion—the erosion by concentrated flow in gullies that appear intermittently during runoff events. Such erosion, however, may be significant, adversely affecting portions of most sloping fields. For this reason, the committee recom-

EXECUTIVE SUMMARY

mends that more accurate and widely applicable measures of ephemeral gully erosion be developed, with a goal of incorporating new concepts and data into future inventories.

Several important refinements are needed in the USLE to further improve its reliability and extend its range of application. Specifically, the committee recommends that accelerated research efforts be directed to

- Ensuring that C factors (representing soil cover and management practices in the USLE) accurately predict the erosion control benefits of crop rotation and cover and management practices;
- Improving the accuracy of field-level determinations of C factors by measuring and/or predicting the actual extent of soil cover provided by crop residues in each year of rotation;
- Developing data and practical methods for incorporating adjustments into the USLE for application to areas of the country where frozen soils, snowmelt, or irrigation substantially alter runoff or erodibility and, subsequently, the estimated rates of erosion using the USLE.

Leveraging Beneficial Uses of the NRI

The committee believes that the NRIs have been underutilized. The USDA is to be commended, however, for steps it has already taken to ensure that the 1982 NRI is accessible and useful to a diversity of analysts and institutions. The USDA's efforts would be more cost-effective and successful if a number of additional steps were taken, including

- Publication of supplemental documentation on both the 1977 and 1982 NRIs addressing in detail survey design and content, methods to test statistical reliability, and caveats associated with common applications of the raw data files;
- Provision of a supplemental tape containing the individual factors of the WEE not recorded on the basic data tapes and the codes needed to cross-reference NRI sample points with other key data sources; at a minimum, these sources should include the SCS Soils-5 file and the hydrogeological data base;
- Publication of a supplemental volume of 1977 and 1982 NRI statistics utilizing a variety of table formats based on ranges of inherent erodibility rather than the Land Capability Class System.

Understanding the Erosion-Productivity Relationship

Previous recommendations will enhance the ability of researchers to more fully employ the NRI as a basic data source. In particular, the NRI can make possible significant methodological and empirical advances in research that will explore the relationship between erosion and productivity. This work is vital, because the susceptibility of soils to erosion-induced damage is variable; at times it is highly variable. The control of erosion on all cropland, where desirable, can be a basic goal; but efforts are most critically needed to identify those soils now incurring accelerating or, perhaps, permanent damage from erosion. New erosion-productivity models, some of which are heavily reliant on the NRI, may soon reach a level of refinement adequate to make possible a long-overdue reassessment of soil loss tolerance limits (T). Better T values should, in turn, directly improve USDA's ability to more accurately classify cropland according to its susceptibility to erosion damage.

Assessing Agricultural Runoff in Relation to Water Quality

Conservation challenges have changed during the last 50 years as a result of progressive technological change in production methods and the change from diversified to specialized agriculture. The change from rotational cropping involving cover crops to monoculture or row crops only, for example, can significantly alter runoff and erosion. Most technological advances to improve yields have had, in different regions and for different crops, both positive and negative effects on erosion-control efforts. By increasing yields and making possible denser plantings, technology has reduced erosion by contributing to more complete soil cover. Some of the same technologies, however, have introduced land use practices that have been very damaging to soils ill-suited to intensive, continuous row-crop production.

Perhaps the most important, relatively new conservation challenge is the need to much more carefully monitor, understand, and mitigate water pollution from agricultural sources. As scientists and farmers improve techniques to keep water on the land, infiltration may enhance the recharging of groundwater. However, such an improvement may increase the opportunity for undesirable chemicals and nutrients to move into groundwater. Technological changes in farming methods involving more widespread, intensive uses of inorganic fertilizers and agricultural chemicals have transformed the types of pollutants that agriculture contributes to surface water and groundwater. For exam-

EXECUTIVE SUMMARY   xix

ple, nitrates have been found in drinking water supplies in several major farming regions, and scattered reports of herbicides and insecticides in groundwater add a new motivation to agricultural management and conservation efforts.

Ample evidence indicates that agricultural sources of water pollution pose significant environmental problems. Better information on the effects of agricultural practices on water quality is essential in devising and implementing strategies for control of these nonpoint sources of pollution. Runoff from the land surface contributes sediments and chemical constituents to water bodies. Both are often considered pollutants. The 1982 NRI contains information on land use and conservation practices that are important in estimating the magnitude of these sources of pollution. The NRI can potentially contribute to research and evaluation of offsite as well as onsite effects of erosion. However, neither in the 1982 NRI nor elsewhere is adequate information yet available to translate with confidence measures of eroding soil and related constituents to impacts on watercourses.

The committee recommends that steps be taken to fully explore ways to more effectively use the 1982 NRI in water quality research. The committee foresees the need—and some opportunities—to incorporate into future inventories additional data designed to enhance the use of the NRI in conjunction with other data sources and models in the evaluation of offsite effects such as sedimentation and pollution caused by erosion of the land.

Magnitude and Scope of Future NRIs

The practical application of future NRIs would be expanded by incorporating improved analytical tools and additional data. The committee has highlighted the need for an improved WEE, incorporation of data and equations to estimate the severity of ephemeral gully erosion on cropland, improved measures relating erosion and productivity, and data necessary to analyze how agricultural management practices affect water quality. In addition to these extensions, the committee believes that future inventories should include appropriate data on federal lands—lands that were not included in the 1982 NRI or earlier surveys.

The committee believes that the costs incurred by including new data in future NRIs might eventually be offset by cost savings made possible by incorporation of new satellite-based sensing technologies into natural resource-monitoring programs. For example, remote sensing information can be used as input data to calibrate and run water quality models. While recognizing the shortcomings of current technologies, the committee anticipates an expansion in the capability to use remote sensing and other technologies in conjunction with computer-based cartographic systems. It might be possible to develop highly refined, analytical capabilities using a combination of field-level surveys such as the NRI to ensure ground truth and new data from satellites that will be available several times each growing season.

New technology should become available to accurately and cost-effectively monitor cropping patterns, rainfall, and erosion occurrences; assess net recharge to groundwater and the effects of agricultural management practices on groundwater quality; and assess the flow of sediment and other constituents off fields and into waterways. Efficient and effective acquisition and use of information from a variety of integrated data bases will require close cooperation among a number of agencies in the federal government.

SOIL CONSERVATION

ASSESSING THE NATIONAL RESOURCES INVENTORY

Volume 1

1
Scope and Content of the 1982 NRI

Between the spring of 1980 and the fall of 1982, the Soil Conservation Service (SCS) of the U.S. Department of Agriculture (USDA), as directed by Congress, conducted the most extensive inventory of land and water resources ever undertaken in the United States. SCS personnel recorded over 70 observations on resource conditions and land use at each of about one million locations across the country. The collected data were entered into computers. The approximate cost of the project was $15 million.

The product of this project is the 1982 National Resources Inventory (NRI): a computerized natural resource data base covering all nonfederal land in the United States. The extensive and in some cases unique contents of the 1982 NRI make it a primary source of data for researchers, government program administrators, and policymakers. Such an extensive survey will not be conducted again until 1987.

The 1982 NRI is the most recent of a series of national resource surveys and inventories performed by the SCS, beginning with the Erosion Reconnaissance Survey of 1936. The first NRI was conducted in 1977 in anticipation of the passage of legislation directing the USDA to evaluate resource conditions and trends. The Soil and Water Resources Conservation Act of 1977 mandated that such an inventory continue on a five-year cycle.

The most important inventory items in earlier surveys were land use, conservation treatment needs, and soil classification. The 1982 NRI and its predecessor, the 1977 NRI, include additional information on prime farmland; the physical and economic potential for future conversion of forest, pasture, and rangeland to cropland uses; incidence of types of wetlands; susceptibility of types of land to flooding; and incidence of soil conservation practices applied on land. New resource items were added for the 1982 inventory, including data on critically eroding areas, riparian vegetation, wildlife habitat diversity, and vegetative cover conditions on rangeland and forestland. (See the Appendix for a formatted listing of the basic data in the 1982 NRI.)

The increased sampling density of the 1982 NRI substantially improves the statistical reliability of the data base and permits, for example, more reliable analysis of smaller geographic areas than was previously possible.

Uses of the 1982 NRI

The Committee on Soil Conservation Needs and Opportunities identified five important applications for the data contained in the 1977 and 1982 NRIs. These uses demonstrate the value of this data source for investigating many aspects of land use and quality, especially soil erosion and conservation. Summaries of the five areas of application follow.

Trend Analysis

Direct comparisons of surveys are not always possible or advisable because of changes in inventory definitions, procedures, sample sizes, and estimation techniques (see Comparing Results from the 1977 and 1982 NRIs). Generally, however, the definitions and procedures pertaining to cropland and its use—a major land use category for erosion studies—are consistent. As a result, data on conversion of land into and out of cropland use between 1977 and 1982 are reliable. Such land use shifts often have a significant effect on soil erosion.

To some extent, trends of this type can also be analyzed using the 1982 NRI alone, because for each sample point land use is identified for the three years prior to sampling. Although rotations were not adjusted to a common 1982 year, Ogg's analysis (1986) of 1982 NRI data suggests that land converted to crop use between 1979 and 1982 generally produced lower yields and eroded at rates higher than average for all cropland in 1982. With some cautions and exceptions, erosion rates reported for major land uses in the inventories can be compared to indicate changes from 1977 to 1982.

Classification of Soils According to Rates of Erosion

Erosion estimates contained in the NRIs permit analysts to identify and investigate areas with particularly high or low erosion rates. As rudimentary as such information may seem, it has not been available until recent years. With the steadily declining buying power of most conservation program budgets, program managers would benefit from more reliable indicators of the most serious problems and from an assessment of the effectiveness of alternative strategies. Reliable indicators are vital to efforts now under way to target government programs to the most immediate resource problems.

Identification of Needs and Opportunities

The NRIs provide a record of the types of conservation measures in use, where they have been adopted, and their effects on erosion. Already, this information has had a notable effect on state and federal conservation policies and programs (American Farmland Trust, 1984).

The NRIs also enable researchers to correlate estimates of rates of erosion observed at thousands of points in the field to other data bases. Thus, they can test hypotheses—within the limitations of these data—about the relationships of erosion to crop yields and sediment loads in waterways, and the effects of alternative conservation strategies.

Data on sheet and rill erosion from the 1977 NRI have been used to investigate the relationship between soil erosion and crop yields across broad geographic areas. Some studies have been made possible by the capability to link NRI data via computer with detailed soil information contained in the computerized SCS soil survey file, Soils-5. (Soils-5, more formally called SCS-SOI-5, is a compilation of soil interpretations data and contains descriptions of the properties of all soils in the United States.)

Investigations of erosion-productivity relationships have led to research on the amount of erosion that soils can withstand before serious and perhaps irreparable damage is done to their capacity to sustain long-term agricultural production (Pierce et al., 1983).

Relating the NRI to Water Quality and Other Environmental Issues

Both inventories appear to have limited applications for direct analysis of water quality problems. But as this area of interest becomes increasingly important in the next decade, new uses might be found for NRI data, possibly in conjunction with new data added to future inventories or as an extension to the 1977 and 1982 inventories. There is a need to enhance future surveys by providing data, where possible, that are relevant to problems of groundwater and surface water contamination by dissolved and particulate substances. Erosion data from the NRIs might be combined with a number of mathematical models developed to analyze water quality problems, particularly those associated with sediment from agricultural land uses (Christensen, 1986).

Inquiries made to USDA by various agencies indicate an interest in other potential research and policy applications of the NRI (G. Nordstrom, SCS, personal communication, 1984). The National Oceanic and Atmospheric Administration plans to use the NRI to study land use trends in coastal areas. The Environmental Protection Agency has expressed interest in using the NRI in its studies of acid rain, nonpoint pollution, pesticide use patterns, and other environmental issues associated with land use patterns. And the U.S. Geological Survey has requested NRI data for use in analyzing observed trends in water quality and for research on nitrate contamination of shallow groundwater.

Comparing Results from the 1977 and 1982 NRIs

Table 1-1 shows the percentage of land in each of the major rural categories for the 1967 Conservation Needs Inventory (CNI) and the 1977 and 1982 NRIs. Forestland increased by 1.5 percent between 1977 and 1982. However, since USDA did not have a uniform definition for that land use until just prior to the 1982 NRI, some land classified as pastureland, rangeland, or other rural land in 1977 was reclassified as forestland in 1982. Additional analysis, especially at the regional or Major Land Resource Area (MLRA) level, could help determine how much of this apparent increase is attributable solely to definitional changes. (MLRAs consist of geographically associated land resource units; groupings by patterns of soil, climate, water resources, land use; and type of farming. A state, for example, might have between 6 and 12 MLRAs.)

The use of improved inventory procedures resulted in an apparent decline of nearly 16 million acres of land in built-up use between 1977 and 1982 (see Table 1-2). Built-up areas include cities, villages, industrial sites, cemeteries, airports, golf courses, and similar areas. This 18 percent change does not reflect an actual decline in built-up

TABLE 1-1 Use of Nonfederal Rural Land (percent), 1967–1982

Land Use	1967[a]	1977[b]	1982[c]
Cropland	29.9	29.5	29.8
Pastureland	35.2[d]	9.5	9.4
Rangeland	—	29.1	28.7
Forestland	30.9	26.4	27.9
Other rural land	4.0	5.5	4.2
Total	100.0	100.0	100.0

[a]Unadjusted data, 1967 CNI.
[b]1977 NRI.
[c]1982 NRI, preliminary data.
[d]Includes rangeland.
SOURCE: 1977, 1982 NRI; 1967 CNI.

acreage during that time but an overestimate of built-up acreage in 1977. Some of this land was reclassified as rural in 1982 as a result of more accurate expansion factors and other improved inventory procedures applied retrospectively to the 1977 survey results. This reclassification and other instances of improved assessments of acreage and classification illustrate the importance of statistical design.

The SCS is conducting a series of studies designed to rectify some of the problems caused by definitional and procedural changes. Until these studies have been completed, NRI users who want to compare data from 1982, 1977, or earlier years might choose to perform preliminary analyses of comparative data for discrete geographic areas that can be confirmed by other sources of information. This analysis can validate the accuracy of NRI-based methods.

TABLE 1-2 Nonfederal Land Uses (million acres), 1967–1982

Land Use	1967[a]	1977[b]	1982[c]
Rural land[d]	1,438	1,401	1,414
Urban, built-up, rural transportation	61	90	74
Small water bodies	7	9	10
Total	1,506	1,500	1,498

[a]Unadjusted data, 1967 CNI.
[b]1977 NRI. NOTE: The decline in rural land from 1967 to 1977 is apparent, not real, and results from improved identifications in the NRI.
[c]1982 NRI, preliminary data.
[d]Includes cropland, pastureland, rangeland, forestland, and minor land cover/uses.
SOURCE: 1977, 1982 NRI; 1967 CNI.

Tool for Research and Policy Analysis

The NRIs have had an important influence on agricultural policy and program administration. The availability of nationally consistent, scientifically derived estimates of soil erosion has made it possible for policy analysts to examine the distribution of national and regional erosion problems more effectively than ever before.

For example, analyses of the 1977 NRI revealed that high rates of soil erosion—and most of the total tonnage that was eroded—were concentrated on a small amount of land. This finding led USDA and others to propose a number of policy and program changes designed to target, or concentrate, federal conservation assistance to specific geographic areas. The inventories have also made possible analyses of the distribution of soil conservation practices nationwide. As a result, policymakers now have a better understanding of the amount of conservation work that needs to be done and the techniques that are most likely to be effective.

Information on erosion and conservation practices contained in the NRIs suggests the potential value of focusing new conservation programs and policies on particular lands. A proposed soil conservation reserve, for example, would offer land rental payments to farmers who voluntarily retire erosion-prone land currently in cultivation. Erosion information has also been used to evaluate so-called sodbuster policies, which propose to eliminate a farmer's eligibility for federal commodity program subsidies if highly erodible land not currently in production is converted to crop uses that are covered by commodity programs. Identification of highly erodible lands—a source of sediment and other pollutants that can affect water quality—may also contribute to selective policies for the control of pollutants. The NRIs have been used to develop and refine land classification schemes that can contribute to administering the conservation reserve and sodbuster proposals.

The committee is convinced that the NRI is an important source of basic information about U.S. agricultural resource conditions and trends. The value and diversity of applications for the NRIs are certain to increase in the future.

Soil conservation and related issues have begun to receive increasing attention from many public interest groups. Increasing interest is also evident within the agricultural community. Polls and surveys of farmers indicate a willingness to accept some forms of mandatory conservation provisions as a requirement for participation in government farm programs (American Farmland Trust, 1984). Public interest in soil

SCOPE AND CONTENT OF THE 1982 NRI

conservation has reinforced efforts within the research community and posed new challenges for refining an understanding of soil erosion processes and their economic, ecological, and environmental consequences. As policymakers are more frequently drawn into discussions on conservation issues, the need for reliable information will increase.

Information from the 1982 NRI

The SCS began releasing data and analyses of the 1982 NRI on a variety of subjects in 1984. Major findings of the 1982 NRI pertaining to land use, soil erosion, and conservation practices that have been reported (see *Soil Conservation: Assessing the National Resources Inventory*, Volume 2) are summarized here. Detailed analyses of NRI data can be found in subsequent sections of this report.

Land Use and Soil Erosion Rates

The acreage in each of eight major uses of nonfederal land is reported in Table 1-3. Of the total of 1.498 billion acres of nonfederal land in 1982,

TABLE 1-3 Use of Nonfederal Rural Land, 1982[a]

Land Use	Acres (millions)
Rural land	
Cropland	421.4
Pastureland	133.3
Rangeland	405.9
Forestland	393.8
Minor land cover/uses[b]	59.6
Subtotal	1,414.0
Urban and built-up land	46.6
Rural transportation	26.9
Small water area	10.1
Total	1,497.6

[a]1982 NRI, preliminary data; excludes Alaska; includes Caribbean.
[b]Farmsteads and ranch headquarters; other land in farms; mines, quarries, and pits; small built-up areas; and other rural lands.

SOURCE: 1982 NRI.

TABLE 1-4 Estimated Average Annual Sheet, Rill, and Wind Erosion (tons/acre·year)[a]

Land Use	Sheet and Rill Erosion	Wind Erosion[b]
Cropland (total)	4.4	3.0
Cropland (cultivated)	4.8	3.3
Pastureland	1.4	0.0
Rangeland	1.4	1.5
Forestland (grazed)	2.3	.1
Forestland (ungrazed)	0.7	0.0

[a]1982 NRI, preliminary data.
[b]Estimates of absolute rates of wind erosion are subject to considerably greater uncertainty than those for sheet and rill erosion (see Chapter 3).
SOURCE: 1982 NRI.

about 94 percent (1.414 billion acres) was classified as rural land. Cropland, rangeland, and forestland are the dominant uses of rural land, each accounting for roughly 30 percent of rural acreage.

Table 1-4 illustrates the substantial differences in erosion rates reported for each use. Cropland has the highest average rates of sheet, rill, and wind erosion. Annual sheet and rill erosion rates on the 421 million acres of cropland average 4.4 tons/acre, almost twice as great as the next highest rate (2.3 tons/acre·year for grazed forestland).

Wind erosion rates for cropland, although subject to much uncertainty, appear to be much greater than the rates for rangeland (see Chapter 3). Erosion rates are yet higher for the 323 million acres of cropland that is cultivated and used for row and close-grown crops or for vegetables, fruit, and other crops. Sheet and rill erosion rates average 4.8 tons/acre·year on cultivated cropland.

These averages obscure enormous variations in erosion rates within each land use (see Chapter 5). But to indicate the subtlety of the erosion process, it is helpful to note that it takes about 100 years, on average, to form 1 inch of soil. The loss of 5 tons of soil from an acre in one year amounts to a layer of soil less than 1/30th of an inch deep, slightly more than the thickness of a dime.

Soil Erosion in Relation to Soil Loss Tolerances

A traditional way to gauge the significance of various rates of erosion is by comparison to the conventional SCS soil loss tolerances (T values)

TABLE 1-5 Distribution of Cropland Acreage by Soil Loss Tolerance

| T value | Cropland | |
(tons/acre·year)	1,000 Acres	Percent
5	300,552	71.4
4	48,355	11.5
3	54,514	12.9
2	15,441	3.6
1	2,540	0.6
Total	421,402	100.0

Average T value (weighted by acreage): 4.55 tons/acre·year

SOURCE: McCormack and Heimlich, 1985.

estimated for most soils in the United States. The T value is defined as the maximum rate of annual soil loss (in tons/acre·year) that will permit crop productivity to be sustained economically and indefinitely. For cropland soils, T values have been estimated to range from 1 to 5 tons/acre·year; 71.4 percent of these soils have been assigned the maximum value of 5 tons/acre·year, and another 11.5 percent have a T value of 4 tons/acre·year. For the generally shallower rangeland soils, T values range from 1 to 3 tons/acre·year (see Table 1-5).

Additional research is needed to increase the reliability and usefulness of the overall concept of T values, as well as the values assigned to specific soils (see Chapter 4). However, conventional T values do convey a sense of which soils are relatively vulnerable to erosion, because values less than T = 5 were assigned to reflect relatively shallow topsoil depth, less favorable geologic material, the relative productivity of topsoil and subsoil, and the historical amount of erosion.

T values were intended to be indicators of the amount of erosion that can be sustained without causing on-farm productivity losses. Additional work is under way to better define the relationship between erosion and potential productivity losses (see Chapter 4). At the same time, recent research (Clark et al., 1985), which includes use of the 1977 and 1982 NRIs, indicates that this concept might be too narrow to represent the most serious consequences of erosion that involve off-farm impairment of water quality in some regions. To the extent that erosion contributes to offsite damages such as water pollution from runoff, for example, soil loss tolerance values in the future might reflect the on-farm and off-farm consequences of erosion.

Soil on this 40 percent slope in the Palouse is eroding at a rate of 300 tons/acre·year (Whitman County, Washington). Severe erosion occurs in early spring when the soil is unprotected. Credit: U.S. Department of Agriculture, Soil Conservation Service.

SCOPE AND CONTENT OF THE 1982 NRI

Some soils are not subject to rates of erosion that impair productivity under routine farming conditions; yet the off-farm effects from relatively low rates of erosion on such soils might still be important in setting tolerances. In such cases, the conventional definition of T values is of limited use, because it is unlikely that erosion could reach levels that threaten on-farm productivity. In other cases the T value needed to protect against off-farm effects might be very high, even higher than the inherent potential for erosion of certain soils. In these cases, the effects on on-farm productivity would be the binding constraint underlying T values.

The rate of erosion from sheet and rill erosion (4.4 tons/acre·year) is roughly equivalent to the average weighted T value of 4.55 tons/acre·year in Table 1-5. Where wind erosion is additive, total average soil loss may exceed these estimated tolerance levels. However, the interrelationship of wind and water erosion is uncertain.

Seventy-five percent of U.S. cropland (315 million acres) was eroding below the tolerance level in 1982 (see Table 1-6). In the Corn Belt about 60 percent of the acreage is estimated to be eroding below 5 tons/acre·year; in Iowa the value is 54 percent. Altogether, cropland eroding at a rate below the T value accounted for over 506 million tons of soil displacement, about 27 percent of the total displacement on all cropland.

Another 13 percent (55 million acres) of the cropland had an erosion rate ranging from the tolerance level to twice the tolerance level (T-2T), averaging 6 tons/acre·year. This land accounted for 330 million tons of erosion, 18 percent of the cropland total. On an additional 51 million acres, cropland erosion measured in the 1982 NRI exceeded assigned T values by a factor of two or more. The average sheet and rill erosion rate on this acreage was 20 tons/acre·year, a rate about four times greater than the national average.

Next to cropland, the most serious sheet and rill erosion problems on nonfederal lands sampled in the NRI are the 353 million acres of rangeland (see Table 1-6). Tolerance values on rangeland soils are generally set at levels that are 1 to 3 tons/acre·year lower than or equal to those for soils in other regions, reflecting the belief among soil scientists that such lands are more sensitive to the consequences of erosion. The erosion rate on about 13 percent of rangeland was at least two times greater than the assigned soil loss tolerance. Erosion problems are less serious on other land uses, primarily because the soil is more protected by the vegetative canopy. Only 9 percent of pastureland (11.5 million acres) and 6 percent of forestland (25 million acres) had an erosion rate greater than the assigned soil loss tolerance.

TABLE 1-6 Sheet and Rill Erosion on Cropland and Rangeland in Relation to Soil Loss Tolerance (T) Values[a]

Acreage	United States				Corn Belt Region				Iowa			
	<T	T-2T	>2T	Total	<T	T-2T	>2T	Total	<T	T-2T	>2T	Total
Cropland												
Acres (millions)	315	55	51	421	69	22	24	115	14	5	7	26
Acres (%)	75	13	12		60	18	20		54	20	27	
Tons (millions)	506	330	1,008	1,843	154	133	535	822	36	36	176	248
Tons (%)	27	18	55	4	19	16	65		14	14	71	
Tons/acre·year	1.6	6	19.8	4.4	2.2	6.2	22.3	7.2	2.5	6.8	25.1	9.4
Rangeland												
Acres (millions)	353	24	29	406	5	0.3	0.2	6	—	—	—	—
Acres (%)	87	6	7		91	5	4					
Tons (millions)	182	63	315	562	3	1	2	7	—	—	—	—
Tons (%)	33	11	56	4	46	18	36					
Tons/acre·year	0.5	2.7	10.9	1.4	0.6	4.8	10.3	1.2	—	—	—	—

[a]The numbers in this table have been rounded for the convenience of the reader. The precise numbers generated from NRI data are the statistical summations of all acreage represented by sampling points; they should be used for further technical analyses.

SOURCE: 1982 NRI.

Concentration of Erosion

One of the major findings of the 1977 NRI was the marked concentration of all forms of erosion on relatively small portions of virtually all land uses. Analyses of erosion data from the 1982 NRI indicate similar phenomena, including variation in the concentration of erosion by land use. Concentration of erosion is most pronounced on cropland, as shown in Table 1-6.

Table 1-7 shows the distribution of sheet and rill erosion grouped into classes according to the rates of erosion. Data in Table 1-7 are confined to the acreage planted to row and close-grown crops, the most important category in terms of extent of acreage, volume of erosion, and intensity of use. Corn, soybeans, cotton, and sorghum comprise the bulk of the row crop acreage. Major close-grown crops include wheat, oats, and barley.

About 325 million acres of cropland were planted in row and close-grown crops in 1982. About 136 million of those acres (42 percent) had a calculated erosion rate of less than 2 tons/acre·year. A total of 107 million acres of land used for row and close-grown crops, 33 percent of the total, eroded at rates between 2 and 5 tons/acre·year. About 75 percent of the most intensively used cropland in the United States (243 million acres) had sheet and rill erosion rates below the average assigned tolerance level for all cropland. This 75 percent of cropland contributed only 30 percent of the total soil eroded (see Table 1-7).

At the other extreme is a relatively small proportion of cropland acreage with high erosion rates, where a disproportionate share of the total soil displacement occurs. On about 7 percent of the land in row and close-grown crops (23 million acres), the average rate of sheet and rill erosion is 15 tons/acre·year or greater. Yet this 7 percent of the acreage accounts for 700,000 tons of soil displacement, or 41 percent of the total tonnage of sheet and rill erosion on cropland used for row and close-grown crops.

Erosion by wind, although estimated with more difficulty and considerably less reliability than sheet and rill erosion (Gillette, 1986), appears to be similarly concentrated (see Table 1-8). Following the pattern observed with sheet and rill erosion, wind erosion is also highly concentrated on a limited portion of cropland. Although the precise figures for gross erosion by wind are in doubt, the relative magnitudes in Table 1-8 are illustrative. Perhaps 5 percent, some 16.8 million acres, of land in row and close-grown crops has very high average annual wind erosion rates (more than 14 tons/acre·year). Yet this 5 percent

TABLE 1-7 Estimated Acreage of Row and Close-Grown Cropland in the United States by Rate of Sheet and Rill Erosion[a]

USLE (Estimated Erosion) (tons/acre·year)	Acreage		Erosion Potential (tons/acre·year)	Erosion (tons/acre·year)	Gross Erosion	
	Million Acres	Percent			Million Tons	Percent
0-<2	136	42	6	1	138	8
2-<5	107	32	14	3	344	21
5-<10	46	14	27	7	316	19
10-<15	15	4	47	12	177	11
15-<20	7	2	63	17	128	8
20-<25	4	1	79	22	93	6
25-<30	3	1	94	27	77	5
30-<40	3	1	115	35	111	7
40-<50	2	0.5	143	45	73	4
50-<75	2	0.5	183	60	104	6
75-<100	0.5	0.2	248	85	43	3
>100	0.4	0.1	375	145	51	3
Total	325	100	1,394	459	1,655	100

[a]The numbers in this table have been rounded for the convenience of the reader. The precise numbers generated from NRI data are the statistical summations of all acreage represented by sampling points; they should be used for further technical analyses.

SOURCE: 1982 NRI.

TABLE 1-8 Estimated Acreage of Row and Close-Grown Cropland in the United States by Estimated Rate of Wind Erosion[a]

Estimated Erosion by Wind	Acreage		Wind Erosion (tons/acre.year)	Gross Erosion	
	Million Acres	Percent		Million Tons	Percent
0–<2	230	71	0.2	50	5
2–<5	42	13	3	139	13
5–<10	26	8	7	187	18
10–<15	11	4	12	139	13
15–<20	5	1	17	83	8
20–<25	3	1	22	67	6
25–<30	2	1	27	57	5
30–<40	2	1	35	81	8
40–<50	1	<1	44	44	4
50–<75	0.9	<1	59	55	5
75–<100	0.3	<1	86	28	3
>100	0.8	<1	143	120	11
Total	324		456	1,050	100

[a]The committee notes that estimates of wind erosion are considerably less reliable than those of sheet and rill erosion. In addition, the numbers in this table have been rounded for the convenience of the reader. The precise numbers generated from NRI data are the statistical summations of all acreage represented by sampling points; they should be used for further technical analyses.

SOURCE: 1982 NRI.

accounts for about half (555 million tons) of the total estimated tonnage of soil displaced through wind erosion in an average year. While the reliability of wind erosion estimates in the NRI can be questioned (see Chapter 3), it is likely that improved estimation procedures and data will affirm the spatial distribution and concentration of erosion of cropland by wind.

In general the geographic locations of significant sheet, rill, and wind erosion problems do not overlap in much of the United States. This results primarily because very different climatic conditions and patterns are associated with different forms of erosion. This may not be the case in some semiarid regions with sparse vegetation and intense rainfalls. However, current information is insufficient to define physical or geographic domains where wind and water erosion processes interact significantly.

Soil Conservation Practices on Cropland

For the 1982 NRI, field personnel were instructed to note the number of conservation practices in use (up to three) at each sample point. Four major cropland erosion control practices—conservation tillage, terracing, contour farming, and stripcropping—were recorded.

(Conservation tillage can be defined as any of a number of tillage systems that reduce loss of soil or water, compared with clean tillage practices that bury all or nearly all crop residues or cover crop. The terms reduced tillage, minimum tillage, and others are often used interchangeably with conservation tillage. No-till is the ultimate form of conservation tillage; the new crop is seeded directly into existing crop residue, cover crop, or sod. Implements are used to make small slits in the soil to place seed. The land is not usually cultivated during crop production.)

Conservation tillage is the dominant conservation practice; it is applied to 49 percent of row crop acreage and 24 percent of total cropland acreage. Other traditional cropland conservation practices occupy a relatively small share of the row crop acreage and the total cropland. Terracing and contour farming, for instance, were reported on 14 and 17 percent, respectively, of row crop acreage and on significantly smaller proportions of the total cropland. Stripcropping was reported on less than 1 percent of cropland.

It should be noted that because more than one practice could be recorded for each sample point, a certain amount of acreage has been

To reduce the threat of erosion and conserve moisture, this new wheat crop is being planted in the residue of the previous wheat crop using a no-till drill (Whitman County, Washington). With this method, erosion will not exceed 5 tons/acre·year. Credit: U.S. Department of Agriculture, Soil Conservation Service.

counted twice. For example, most terraces are constructed along the contours of a field, more or less dictating the use of contour farming within each terrace interval. Thus, much of the acreage reported for terrace systems is also counted for contour farming. Less than 64 million acres were protected either by terraces or contour farming.

The same double-counting is likely with other practices. Thus, the sum of acreages for two or more individual practices, and the total acreage for all four, is often an overestimate of the total acreage treated. In 1982 at least one of these four conservation practices was used on approximately 167 million acres. Conservation practices are described in detail in Chapter 5.

2

Improving Federal Resource Assessment Efforts

Major surveys on natural resource use patterns and conservation needs have been conducted by the USDA nearly every five years for several decades. The NRI series has been expanded to include new data and refined to include more precise definitions and sampling procedures. In addition to improving the statistical validity of the sampling, a major recent methodological advance was use of the Universal Soil Loss Equation (USLE) and Wind Erosion Equation (WEE) to collect information about erosion conditions and conservation practices on U.S. nonfederal lands in a consistent, reliable manner.

The 1982 NRI represents a significant improvement over earlier resource inventories. Field procedures; sampling methods; and density, quality control, data accuracy, presentation, and documentation were of a higher quality in 1982 than in 1977. The SCS has consistently upgraded its resource inventories over the past 25 years.

To some extent, successive refinements in areas including inventory design, procedures, and definitions have made direct comparisons among inventories, and subsequently assessment of resource trends over time, difficult. In general, however, the committee believes that the SCS has struck a favorable balance between the often contradictory goals of providing consistent inventory data over time and improving the procedures and reliability of successive inventories.

There will always be a need for resource inventories. The U.S. Congress has already expressed interest in a program of periodic resource inventory and appraisal in the Soil and Water Resources Conservation

Act of 1977, which calls for another inventory in 1987. In this chapter the committee identifies ways to improve future inventories.

The next generation of NRIs will be influenced by several developments. As scientific understanding and public concern about resources and environmental quality evolve, it is likely that the types of resource data collected in successive inventories will change. For example, the committee believes that information directly relevant to water pollution problems derived from nonpoint sources will become an important goal of future resource assessments.

The scale of these future inventories—the number of sample points and the number of observations made at each point—will be conditioned by cost considerations. In the past few years, USDA officials have expressed a desire to lower the cost of the NRIs by reducing their scale and by conducting them every 10 instead of every 5 years, supplemented perhaps with smaller special surveys. Many experts believe that future advances in sensory technology can replace at least some of the costly on-the-ground data collection characteristic of past inventories. In addition, supplemental funds might be sought from states, particularly if a state is interested and willing to support sampling at smaller units, at the county level, for example.

Government agencies should begin to take these developments into account now if resource inventories are to be even more reliable, useful, affordable, and timely in the future. As the principal agency involved in the planning and execution of NRIs, the SCS will have to upgrade the status, financial support, and procedures of its inventory and monitoring functions. A more formal degree of interagency coordination and planning will be needed at the earliest stage of inventory development to ensure that the next generation of resource inventories serves even broader purposes than the 1982 NRI. Federal agencies involved in resource management will need to experiment with such new survey techniques as remote sensing for the NRI and to refine established procedures such as use of the USLE. Completed inventories will have to be compiled and distributed in forms that reflect the most up-to-date concepts of land classification and at the same time can be used interactively with other data bases.

The committee stresses that all of these steps will require a sustained and systematic effort over many years. A comprehensive review of federal resource and environmental inventories is beyond the scope of this report. However, the committee has identified the need to improve the overall planning and design of future NRIs in four areas: (1) compilation and dissemination of data, (2) use of remote sensing and related technologies, (3) coverage of federally owned land, and (4) coverage of parameters related to water quality.

IMPROVING FEDERAL RESOURCE ASSESSMENT EFFORTS

Compilation and Dissemination of NRI Data

The committee sees a need to improve several aspects of NRI data compilation and dissemination: the inventory and monitoring function of the SCS, statistical documentation, presentation of published data, and distribution of NRI computer tapes.

Upgrading Inventory and Monitoring Functions of the SCS

Given the massive scale of every phase of the 1982 NRI, the SCS has performed in a timely and exemplary manner in completing the survey, checking its quality, and releasing inventory results. The committee believes, however, that certain delays in processing NRI data might have been avoided if the inventory and monitoring functions within the agency—at its national headquarters, in particular—had been given higher levels of support, permitting larger staff or greater latitude in seeking assistance outside the agency.

Given the scale of the NRIs and the need to provide inventory results as soon as possible to the field staff of the SCS and other users, SCS should periodically increase the size and financial support level of its inventory and monitoring staff within the national headquarters.

The committee believes that it would be especially useful to increase the number of technical staff with training in statistics, computer science, computer-aided mapping, and remote sensing to permit more timely completion of quality control checks, inventory documentation, and analyses.

The committee recognizes that budgetary constraints and the desire to increase field staff have led to a reduction in staff at the SCS national headquarters in recent years. The committee believes, however, that inventory and monitoring will become central functions of the agency in the future. Therefore, an upgrading of the inventory and monitoring functions should begin immediately to ensure that the many statistical and analytical tasks remaining for the 1982 NRI are completed as soon as possible and to begin preparing for subsequent inventories.

Statistical Documentation

The experiences of researchers working with NRI data have shown that better basic documentation on the 1982 NRI is needed (Brown, 1983; U.S. Congress, House, 1981). Documentation is needed for both the 1982 and 1977 NRIs. It is important that SCS take the lead in fostering enhanced sophistication in the use of NRI survey data.

A diversity of resource issues can be addressed with NRI data. The full benefits to society of additional applications of NRI data will come about in conjunction with the increasing range, complexity, and geographic specificity of NRI-based analyses. Such applications, however, will stretch—and could exceed—the statistical reliability of the underlying NRI data unless steps are taken to inform NRI users about the appropriate statistical confidence limits that should be ascribed to various aggregations of NRI sample points.

The committee believes that the agency's desire to provide the highest quality NRI data might have led to excessive hesitancy in applying data to further analyses that might involve questions of statistical reliability. Rather than forego these types of analyses, the SCS should take steps to objectively quantify the degree of statistical reliability associated with them. Often analysts can gain insights by assessing the relative values and distribution of values of certain NRI parameters in a given geographic area or by comparing regions, even if the absolute value of the parameter for some primary sampling units may be variable and of limited statistical reliability at a particular level of aggregation.

Documented procedures for testing the statistical reliability of most NRI data points and applications are readily available, primarily through calculation of confidence intervals. However, the committee strongly urges SCS to place a high priority on the development of appropriate statistical procedures and other tools needed to carry out more sophisticated tests of significance and reliability.

Procedures addressing statistical reliability tests for the 1982 and 1977 NRIs should be developed and disseminated to users as a first order of priority in conjunction with new applications of the survey data. Reliability issues that arise from joint applications of the 1982 and 1977 NRIs in time series studies also deserve special technical attention.

To address documentation needs in past NRIs and to provide an optimal basis for structuring and reporting results of future NRIs, the committee recommends the following:

Documentation of NRI computer tape formats and contents should be accompanied by a detailed, technical statistical guide.

Such a statistical guide should accomplish the following:

- Describe the sampling method and the way expansion factors are calculated and used in extrapolating from the 3 percent NRI sample of data points to nationwide coverage of nonfederal lands;
- Include instructions for the calculation of confidence intervals for any specific data element at various levels of aggregation;
- Offer step-by-step recommendations showing how analysts can

IMPROVING FEDERAL RESOURCE ASSESSMENT EFFORTS

incorporate statistical reliability checks into specific applications of the raw data.

Because NRI data will be applied to a diversity of research studies requiring varying degrees of accuracy, instructions are needed to build specific criteria for confidence intervals into modeling exercises. SCS should develop, refine, apply, and explain such criteria in the course of compiling the special publication of 1977 and 1982 NRI results described above.

SCS should designate an individual or office within the agency that can offer assistance to outside users in resolving technical questions related to statistical reliability of NRI applications. Assistance in responding to such inquiries could also be provided by the Iowa State University Statistical Laboratory at Ames and SCS National Technical Centers.

SCS should publish a special report that describes applications of the NRI surveys and earlier inventories to time series analyses of conservation needs and accomplishments and land resource use patterns. Such a report would address the implications for time series studies of the changing definitions, statistical design, and data elements in these surveys. Special attention should be directed toward ways to use the 1982 NRI for retrospective studies of conservation needs and accomplishments.

Presentation of Published Data

For many NRI users, tabular materials published in reports or other forms by SCS are the only NRI data consulted, used, or needed. Traditionally, published data pertaining to soil erosion have been organized using the Land Capability Class System (LCCS). Using this system, Roman numerals I through VIII indicate progressively greater limitations and narrower choices for practical use of the land. For example, class I soils have few limitations that restrict their use; class VIII soils have limitations that preclude their use for commercial plants and restrict use to recreation, wildlife habitat, water supply, or esthetic purposes. The additional letters w, s, c, and e are designated subclasses that indicate, respectively, whether the problem is caused by wetness; shallowness, drought, or stoniness; climate; or erodibility.

In focusing on erosion and its potential effects, tables that report erosion rates by land capability class are of more limited value than classifications now available from the NRI based on characteristics specifically related to potential erodibility. Reporting in tables should reflect the appropriate focus and scale. Thus, the LCCS, reflecting a number of influences, may be the appropriate reporting form—for example, where data refer to the rate of adoption and effect of conservation practices, especially practices routinely recommended specifically for soils within certain land capability class and subclass units.

The SCS should publish tables of 1982 NRI data that report erosion rates, quantities of eroded soil, effectiveness of erosion control practices, and other measures of erosion control needs and accomplishments according to the ranges of inherent erodibility of land or susceptibility to erosion-induced productivity losses.

Conservation needs and the effectiveness of conservation practices can be analyzed and understood most reliably through use of the USLE and its individual factors. Based on runoff plot experiments, the USLE is used to predict the longtime average soil losses in runoff from specific field areas in specified cropping and management systems. (The WEE similarly was developed as a method for estimating the potential for wind erosion in the field.)

Reports containing 1982 NRI results with tables based primarily on the LCCS should be supplemented with additional tabular material. Supplemental 1982 NRI reports and subsequent surveys and analyses should include tabulations reporting erosion rates and gross erosion—estimated by the USLE, where appropriate, and an improved WEE—by ranges of erosion and by ranges of the inherent potential for erosion measured by the RKLS-product values from the USLE (see Chapter 3 for discussion of RKLS). Providing 1982 NRI data in tables using the LCCS and RKLS-based classification systems will facilitate an orderly transition to the use of appropriate alternative land classification schemes, a task the committee sees as essential and now technically feasible.

The SCS should encourage the transition within the conservation and land use research community toward improved presentation of NRI survey results by publishing a special statistical report that includes data from both the 1977 and 1982 NRIs.

The SCS report should include discussion of several issues that will be raised in subsequent chapters of this report, including statistical reliability, the effects of definitional and other changes on time series use of NRI data, and the relationship between alternative land classification systems. The report should also include the new tables described above, in which acreage is partitioned according to ranges of inherent erodibility. The report should discuss and contrast various land classification schemes, with emphasis on the most appropriate and inappropriate applications of each. The analysis and experience gained in producing this special report might provide further insights for structuring and reporting results of future inventories.

IMPROVING FEDERAL RESOURCE ASSESSMENT EFFORTS 25

In developing the content of statistical reports on future NRIs, the agency should seek opportunities to extend NRI data to users involved in water quality research and planning and other aspects of land use analysis outside the basic mission of the SCS.

Special aggregations of NRI data by watersheds or water resource regions would be of interest to water quality researchers. To improve such aggregations, additional water resource area geographic codes could be added to future NRIs. In addition to these broad areas, the specific problems of salinity in land and water have been called to the attention of the committee and warrant consideration in future resource inventories.

Other agencies might agree to assist in supporting and executing future NRIs if the USDA can demonstrate a willingness and capability to tailor publications, data, raw data, tapes, and perhaps surveys to the study of some of the analytical problems confronting other agencies.

Distribution of NRI Computer Tapes

Efforts by the SCS to disseminate data tapes of the 1982 NRI have been exemplary. The potential value of these inventories to society depends on the widespread use of their basic data in analyzing conservation problems and evaluating alternative solutions. Wide dissemination of the NRI data tapes to land grant universities, experiment stations, Agricultural Research Service (ARS) research sites, and private institutions helps to ensure that the full potential of the NRI is realized.

The committee believes that the positive experiences to date from outside use of the NRI confirm that further steps are now warranted to make NRI data tapes available to a wider range of users. Several relatively simple steps should be taken to broadly extend the use and dissemination of NRI survey results and coordinate the work of independent analysts using the NRI series.

In NRI data tapes, the SCS should include for each sample point the soil series identification code needed to link the NRI data file to Soils-5, the cooperative, computerized soil survey file.

Soils-5 is an important but limited source of additional data on the physical and chemical properties of individual soil series. Linkage involving soil properties, soil mapping units (where available), and NRI data points can be helpful in carrying out many types of research tasks, including quantification of the relationships among erosion, land use, and crop productivity.

If demand warrants, the USDA should consider making available for researchers a combined computer tape that includes the following for each point in the 1977 and 1982 NRIs:

- Identification codes for the primary sampling unit, including the soil series code;
- Pertinent data from the Soils-5 file for each point;
- Individual data elements on land use, erosion, and conservation practices in 1982 and 1977 for those points covered by the 1977 NRI;
- Individual factors of the WEE.

The SCS should have the capability to provide, upon request, these data tapes for any user in the United States and its common geographic divisions. This capability would aid in making cost-effective responses to requests for specific data in defined geographic regions.

By developing this capability and responding to requests, the USDA can guarantee that NRI data are widely and properly used by analysts. In particular, SCS could recommend against the release of data of questionable statistical reliability—detailed land use data by county, for example—and ensure that data users understand the definitions and statistical reliability of specific data.

The USDA should implement a means to provide researchers with geographic coordinates for the location of NRI sample points to facilitate linkages with remote sensing and computer-based geographic information systems.

Progress in mapping, education, and analysis has been made by linking NRI data in selected states to geographic information systems. Johannsen (1986) has described the potential uses for the NRI in state and local decision making in Missouri, for example. Concerns regarding the need to preserve the confidentiality of landowners have been successfully resolved in situations involving the NRI and other surveys. These experiences should be drawn upon in developing strategies for sharing geographic coordinate information with other researchers who pursue sound research objectives and agree to abide by procedures designed to protect confidentiality.

SCS should continue to develop and implement strategies to improve the accessibility and application of the NRI series. This task could be assigned to an existing group or to a new working group composed of federal agencies involved with resource management and environmental quality. Representatives of state governments, the private sector, and academia should be involved.

Use of Sensing Technologies in Future NRIs

Improvements in most existing natural resource data bases, including the NRI, generally would entail additional costs. Current federal

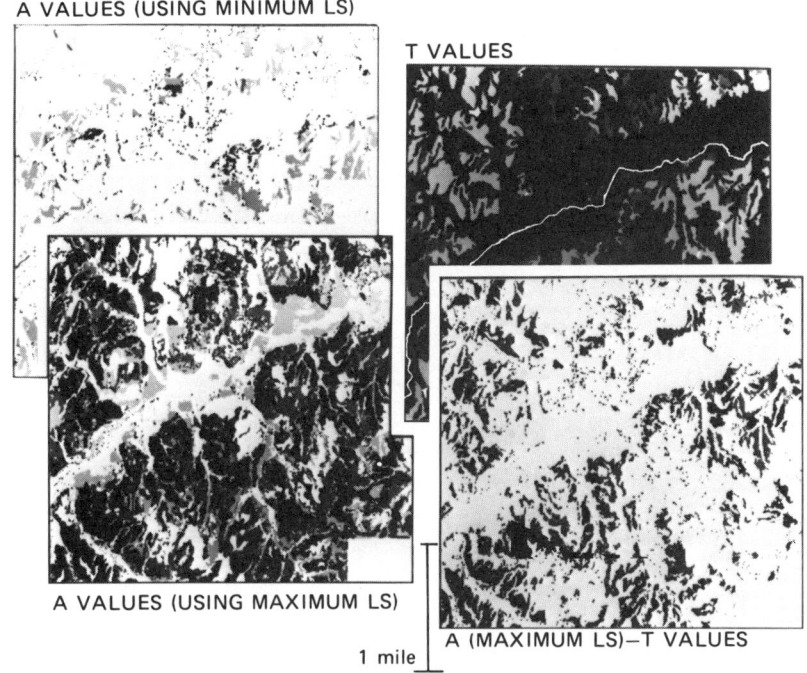

The four images were derived from a combination of aerial photography and digitized soil survey maps of the Little Washita River watershed near Oklahoma City, Oklahoma, to evaluate nonpoint pollution. The A values represent the average annual soil loss per acre predicted by the Universal Soil Loss Equation (A = RKLSCP). In these images the values for L (slope length) and S (slope steepness) vary. T is the soil loss tolerance value, and A − T shows the soil loss beyond tolerance. (The white line on the T-value map indicates the river channel; little erosion is occurring in the floodplain.) White indicates a value of zero; values increase from light to dark shades of grey, the darkest indicating most severe erosion. Resolution is 30 meters. Credit: R. E. Pelletier, National Aeronautics and Space Administration.

expenditures for data collection are low. New sensory technologies, including remote sensing, in time might offset budget constraints for some resource inventory applications by replacing costly on-the-ground surveys.

The committee is aware that promotion of remote sensing has exceeded the capacity of the new technology to define those features on the ground at the level of specificity defined by particular users. Never-

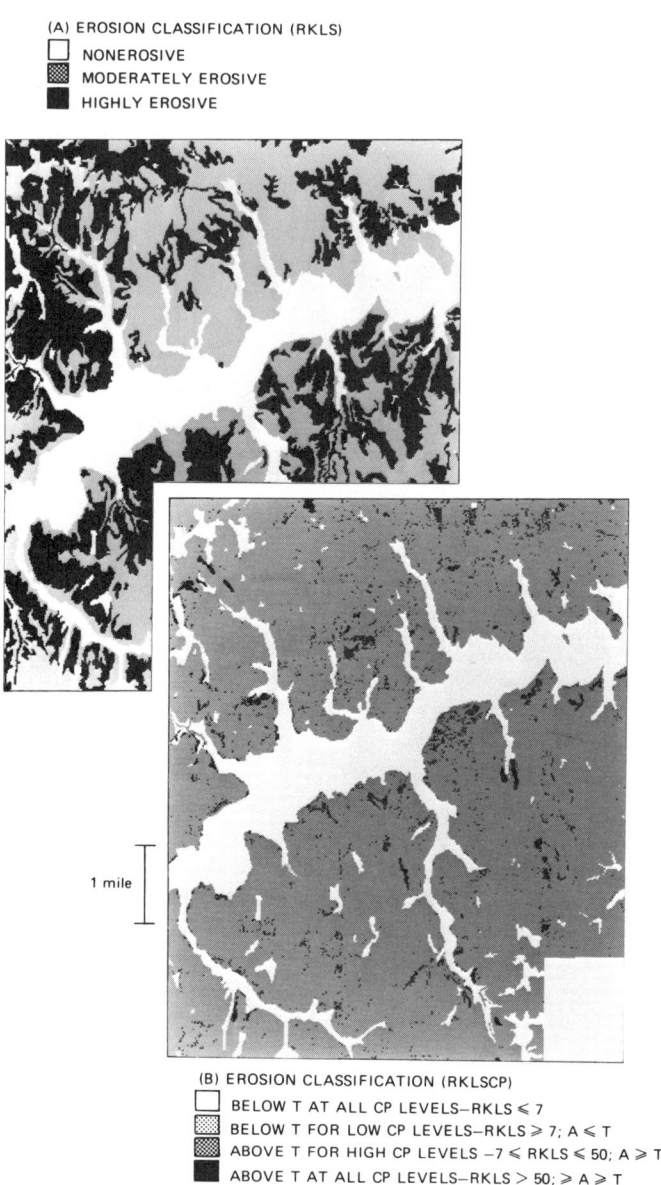

Images derived by using remotely sensed data for the cover and management factor (C) with additional digitized maps show alternative erosion classifications based on (A), the physical (RKLS) components of the equation only, or (B), the conventional equation (RKLSCP) classification. Credit: R. E. Pelletier, National Aeronautics and Space Administration.

theless, improvements in technologies for both sensing and data handling continue to hold promise and demand coordination in development, evaluation, and use. The committee's comments are directed toward this promise and to the necessity of taking a national view of resource inventories.

No single federal agency will be able to take full advantage of the new developments anticipated from subsequent generations of civilian remote sensing technology. Ideally, however, coordination among the many resource inventory activities of the federal government covering the country as a whole could be enhanced to take advantage of remote sensing, computer-assisted mapping, and related survey methods without a major overlap in effort.

Integration of data collection among the federal agencies is difficult. Different information needs and survey procedures and a reluctance to share data collection functions are common barriers to coordination. Yet, because sensing technologies have a variety of applications and hold promise for upgrading current resource assessment activities at a reasonable cost, the committee believes that federal coordination is essential. Because of the potential for public benefit, it might be appropriate for defense agencies to reevaluate policies that in the past have precluded civilian applications of sensing technologies and other capabilities developed for military uses.

The NRI, because of its scope and cost, is an ideal way to pursue methods of replacing or supplementing on-the-ground procedures with sensing technologies. The SCS has made some use of remote sensing and has investigated ways of integrating it into the NRIs. However, the agency could benefit by a more concerted and fully funded effort.

Innovative steps have already been taken to combine remote sensing, various forms of aerial photography, and NRI data. These combined techniques are likely to be of particular value at the state and local levels and will enhance progress in mapping, conservation program administration, and other analytic activities. Sensing technologies can also expand the coverage and usefulness of NRI data to include water quality and groundwater use assessments.

The committee believes that a fundamental transition in the basic method of collecting much of the NRI-type data may occur within decades. Future inventories will be based on remotely sensed data and coded to computer-based cartographic systems that will be equipped with highly refined analytic capabilities. The resource data system of the future, like contemporary NRIs, will be dependent on periodic field surveys to provide ground verifications. Field surveys will be pivotal in verifying and refining the analytic tools used in translating remotely sensed images and high-altitude aerial photographs into geographic information systems and models.

Ad hoc committees and special groups have been convened within and among agencies at times to review topics specific to special areas, digital cartography or agency-specific remote sensing research needs, for example. A formal group that provides a continuing assessment of technologies and applications would benefit the design and management of future resource data systems.

The committee recommends that a federal interagency working group, coordinated by the Office of Science and Technology Policy or another appropriate agency, be formed to formally evaluate potential applications of existing and future sensing technologies to natural resource inventories. The working group should proactively determine types of new sensing technologies that could be developed to support USDA in carrying out resource management responsibilities. The working group should possess the scientific capability to assess and exploit technological advances occurring in remote sensing, computer-assisted mapping, and other survey and analytic methods. The committee recommends that the appropriate agencies of the USDA, the Department of the Interior, the Environmental Protection Agency, and the National Oceanic and Atmospheric Administration be represented in this working group. In addition, appropriate agencies of the Department of Defense should be represented on an ad hoc basis to provide information about sensing technologies that could be useful for future resource assessments.

Expansion of NRI Coverage to Federal Lands

The U.S. government owns large tracts of land, particularly in the western United States. A majority of the acreage of some states is federally owned.

Traditionally, SCS inventories have not included federally owned lands. However, surveys are needed of resource conditions and trends on federal lands, substantial portions of which are in grazing and forestry uses (Sampson, 1986; Renard, 1986). The absence of data on erosion, range conditions, and conservation needs for large portions of certain states, particularly in the West, compromises the NRI as a source of information for public and private management of land resources.

A committee representing the Forest Service, the Bureau of Land Management within the Department of the Interior, the U.S. Geological Survey, the Department of Defense, the Department of Energy, and other relevant federal agencies should be formed to evaluate the quality of current information about resource conditions and trends on federally owned land.

The primary purpose of this evaluation would be to determine if a supplemental inventory effort is warranted. The evaluation would identify the resources and parameters that should be considered in the design of a new inventory and the methods and institutional responsibilities that will be required. The committee strongly recommends inclusion of information on rangeland conditions and soil erosion estimates in any inventory of federally owned lands. This proposed committee should be coordinated with the interagency working group that assesses applications of sensing technology to NRI survey needs.

The supplemental inventory should be designed and conducted to be consistent with the 1982 NRI. Parameters collected on federal lands will include some of those collected for all sample points on private lands. Some data, however, will not be appropriate for the majority of federal lands, for example, cropping history and conservation practices. Other data not collected from private lands might be needed to characterize the uses and conservation needs of federal lands, such as whether lands managed for multiple uses are subject to state or federal conservation programs or initiatives. The appropriate level of detail and density of geographic sampling are yet to be determined. The detail achieved in the 1982 NRI, upon analysis, might not be necessary.

Efficient natural resource data collection, compilation, and dissemination at the federal level will require the cooperation and collaboration of many agencies that are now responsible for specific surveys and data bases. To the user seeking comprehensive data, shortcomings related to coverage in USDA's NRI series are analogous to those in other natural resource data bases.

Inclusion of Data Related to Water Quality

The effects of agricultural production practices and other human activities on water quality are viewed as major environmental problems in many parts of the country. The committee believes that agriculture's potential contribution to improved groundwater and surface water quality will be an important consideration in future research, farm policies, and regulatory and educational programs. The increasing emphasis on water quality reflects two reinforcing trends.

First, substantial progress has been made in identifying major point sources of water pollution and initiating control. Agricultural operations in many regions are now recognized as the most important nonpoint sources of water pollutants. Second, the potential hazards of agricultural pollutants to health and ecological systems have changed dramatically in the last 20 years with increasing rates of application of fertilizers and changes in the toxic properties and use patterns of crop protection chemicals. Runoff from cultivated cropland into surface waters often transports pesticides and fertilizers, in addition to sediments. Water percolating from surface soil into aquifers can carry pollutants into underground water supplies.

Currently, detailed monitoring data are limited for agricultural pollutants in surface water and groundwater. Data are particularly sparse for pesticides. Whether extremely low levels of contamination are of toxicological significance, steps to minimize the flow of agricultural pollutants into water are viewed as imperative and prudent.

Devising long-term solutions for reducing the volume and hazards of these water pollutants will require better information on the pollutant loadings and hydrology of agricultural watersheds. Such information is not collected as part of the NRI, although a number of U.S. Geological Survey water quality network stations gather some information of this type.

Insights can be gained from research on the effects of agricultural management practices on water quality. The Iowa Geological Survey (1984), for example, has combined sophisticated hydrologic monitoring studies with assessments of on-farm cropping, conservation, and agronomic management practices in several parts of the state. These studies have documented many complex interactions between conservation and other management practices. For example, some conservation practices that help control soil erosion may increase the potential hazard for surface water contamination by fertilizers and pesticides.

Contamination of groundwater by fertilizers or herbicides has been documented in areas of more permeable soils and rock formations (Iowa Geological Survey, 1984). Runoff of water carrying sediment, nutrients, herbicides, and other materials from agricultural land has a major influence on the quality of surface water in many regions of the country (Baker, 1984; Gianessi and Peskin, 1981). In the future, conservation systems on the farm may be tailored to the dual goals of controlling erosion and mitigating the potential impact of dissolved constituents and sediments on the quality of surface water and groundwater.

To achieve these objectives, continuing study is needed to develop or to improve, test, and evaluate models of water quality. This process itself helps to identify the types of field data that may be useful for monitoring those activities that influence water quality. In turn, data

IMPROVING FEDERAL RESOURCE ASSESSMENT EFFORTS

from the NRI can contribute to model development and evaluation and, subsequently, monitoring.

After giving consideration to the costs of data collection, other information sources, and the competing uses of natural resource inventory funds, the following kinds of data should be evaluated for potential inclusion in future inventories:

- Additional data needed to estimate sediment loads to streams from estimates of erosion; the parameters needed to calculate the ratio of soil leaving a farm field and entering waterways for sheet, rill, and ephemeral gully erosion; and the distribution of sediment offsite, particularly related to lakes and reservoirs;
- Types, sources, and magnitudes of nonpoint pollution from such nonagricultural sources as uncovered landfills, construction sites, and surface mining operations;
- Deposition of wind-borne soil particles into surface water;
- Farming activities that may be related to specific groundwater pollutants.

There are practical and financial limits to the amount of data that might profitably be included in future NRIs. Yet the NRI already includes observations relevant to water quality, and in many cases no practical and accurate survey technique exists for broad-scale inventories such as the NRI. In view of the importance of water quality problems nationwide, the committee believes that it is useful to evaluate further ways in which the NRI can be effectively related to the federal network of water quality data collection, and vice versa. Future NRIs can play an important role in a broader, better coordinated federal data collection effort to measure and analyze water quality trends. Other agencies may agree to assist in supporting future NRIs if the USDA can demonstrate a willingness and capability to include new data.

It would be logical to incorporate into future NRIs new survey methods for assessing nonpoint water pollution problems and their causes. The committee believes that the collection of data to assist in reliably assessing the impact of agricultural practices on water quality should be the next extension of the NRI data series.

3

The Measures of Soil Erosion

It is possible today to locate the most highly erodible croplands in most regions of the country. Likewise, land not subject to serious erosion under any management system also can be readily identified. The effect of alternative conservation practices on erosion losses can be accurately estimated for a diversity of field conditions. Although the effects of the concentrated flow of water on erosion and sediment yield are not well established, many of the basic physical factors contributing to erosion and accounting for erosion control are well known. These analytical capabilities—made possible by the development of physically based erosion equations—have revolutionized conservation planning and program administration.

The Universal Soil Loss Equation (USLE) is used to estimate the long-term average amount of soil displaced by the forces of rainfall and water runoff along a specified slope. Similarly, an equation has been developed that represents soil loss caused by wind; however, the Wind Erosion Equation (WEE) is far less accurate than the USLE. Currently, there is no widely accepted, practical method for estimating another form of erosion known as ephemeral gully erosion. The topography of most fields causes runoff to collect and concentrate in a few major natural waterways or swales before leaving the fields (Foster, 1982; Thorne, 1984). These features are often ephemeral, and the erosion that occurs in them can be called ephemeral gully, concentrated flow, or megarill erosion. Ephemeral gully areas within fields are plowed in and tilled across annually, in contrast to the permanency of classical gullies

(Foster, 1986). (Ephemeral gully erosion is discussed in detail in the final section of this chapter.)

NRI Estimates of Sheet and Rill Erosion: The USLE

Developed in the late 1950s, the USLE is designed to predict long-term average soil losses through sheet and rill erosion from specific land areas under specified cropping and management systems. In a sense, *soil loss* is a misnomer; *movement* or *displacement* are better terms. Eroding soil is never lost in the sense of disappearing. Often it is merely moved from one part of a field to another, to be deposited in low-lying parts of the landscape. In other cases, soil is moved from the land surface and transported in streams and rivers. (Soil eroding off and down a sloping field, however, is "lost" from its point of origin.)

USLE estimates, commonly expressed in terms of tons/acre·year, do not accurately represent the amount of soil that leaves a field, enters a body of water, or otherwise contributes to offsite erosion damage.

Heavy sheet and rill erosion on sloping cropland, probably exceeding 30 tons/acre·year (Pottawattamie County, Iowa). Severe sheet and rill erosion is limited to a relatively small proportion of cropland in the United States. For example, approximately 7 percent of the land in row and close-grown crops accounts for 41 percent of the total tonnage of sheet and rill erosion on land in that use. Credit: U.S. Department of Agriculture, Soil Conservation Service.

(Models designed for this purpose are in use, as noted in Chapter 2.) Nor do erosion rates estimated by the USLE necessarily correspond directly to the severity of onsite damages to land productivity, since such factors as soil depth and subsoil quality are also important determinants of a soil's vulnerability to erosion.

Rather, the USLE estimates the long-term average amount of soil displaced by the forces of rainfall and water runoff along a specified slope. The segment represented may be on or off the field, at the bottom of a hill, in a terrace channel, or in a natural depression along a slope.

The form of the equation is:

$$A = RKLSCP$$

where A is the computed soil loss per unit area over a specified time; it is usually expressed as tons/acre·year. The factors R, K, and S reflect characteristics of climate and land that generally cannot be modified by human activity designed to influence erosion rates: amount and intensity of rainfall (R), soil erodibility (K), and steepness of field slope (S). The factor reflecting length of slope (L) can be reduced by installing terraces, which effectively break the naturally occurring slope length into smaller segments. (The effective slope length can also be shortened by stripcropping and grassed waterways, but this is reflected in the P factor.) As noted in Chapter 1, however, terraces are in use on limited acreage nationwide. The remaining two factors reflect the effects of human activities on erosion rates: soil cover and management practices (C) and supporting conservation practices (P).

R is a numerical indicator of the erosive forces of rainfall and runoff, developed from rainfall data averaged over several decades. The R factor is the single most influential factor in the equation. Its value varies from 550 along parts of the high rainfall Gulf Coast to 20 in the arid West. The values for R are generally less reliable for western regions in any given year. This occurs because data from the West were insufficient to develop the equation, and rainfall there is intermittent and often torrential, causing average annual sheet and rill erosion estimates to vary greatly in those areas. For these and other reasons, concern has been expressed that the equation is not suited to evaluate potential erosion from rangelands (Society for Range Management, 1985). The committee has not dealt specifically with rangeland in this report.

Other sources of runoff are also important in parts of the West. For the 1982 NRI, modifications of the R-factor values were made for sample points on frozen soils in the Pacific Northwest.

K reflects the inherent susceptibility of a soil to erode, if it is barren of crop cover or residue and exposed to rainfall and runoff. The values for K (ranging from 0.7 for highly erodible soils to 0.2 for soils resistant to erosion) are a function of texture (the percentages of sand, silt, and clay-sized particles), organic matter content, physical structure, and permeability to water.

L and S factors represent the effects of slope length and steepness, respectively. The erosive force of water runoff increases as slope steepness and length increase; the values of these factors increase accordingly. In field applications, including the 1982 NRI, these factors are combined into a single LS factor according to procedures in SCS technical guides or Agriculture Handbook No. 537 (Wischmeier and Smith, 1978).

Determination of a value for LS in the field can be an imprecise exercise. If a prevailing slope can be identified, the surveyor can usually estimate slope steepness fairly accurately with a clinometer, an instrument designed to measure angles of elevation or inclination. Slope length is a parameter that is difficult to evaluate. Once the vector is selected, however, slope length can be determined by measuring with a tape or by pacing. If slopes vary greatly in a field, it is difficult to determine which is the prevailing slope. An average slope steepness is often assigned to a field after clinometer readings have been made for several slopes. Similar judgments must be made to estimate slope length.

The product of the four factors, RKLS, yields an estimate of the average annual sheet and rill erosion (in tons) expected if an area of land were tilled continuously up and down any prevailing slope and kept barren of vegetation. These conditions correspond to a C value of 1 and a P value of 1. The value for the RKLS product thus represents a soil's inherent potential for sheet and rill erosion. Any erosion control practices reduce soil loss on a particular field by lowering either or both of the C or P values below 1.

C is the vegetative cover and management factor. Values for the C factor are multiplied by the product RKLS to represent the reduction from inherent erodibility brought about by cropping sequences, tillage practices, and plant residues on the soil surface. For example, a C-factor value of 0.30—the average value for cropland in the 1982 NRI—means that vegetative cover and management reduced sheet and rill erosion on cropland to 30 percent of its inherent potential rate, estimated by the product RKLS. C-factor values represent the ratio of soil loss from land cropped under specified conditions to the corresponding loss from clean-tilled, continuous fallow. Values for the C factor range from as low as 0.003 for a dense vegetative cover such as permanent, high-quality pasture to 0.7 for crops that produce very small amounts of residue—cotton, for example—and fields that are extensively tilled (see the boxed article Key Role of C Factors in Controlling Erosion).

The factor P reflects the erosion control effects of such supporting conservation practices as contouring, stripcropping, and terracing. These practices break up the lengths of downslope segments traveled by runoff water into shorter segments, thus limiting the volume and velocity of moving water. As a result, less soil is displaced and transported. The P factor is the ratio of soil loss with a specific support practice or practices to the corresponding loss with up-and-down slope cultivation. Values for the P factor are multiplied in the equation in the same manner as C values. A value of 0.91—the average value for P on cropland in the 1982 NRI—means that the estimated sheet and rill erosion would be reduced 9 percent below the rate that would occur without the supporting practice.

NRI Findings On Average USLE-Factor Values

National average 1982 NRI values for the RKLS product (inherent potential for sheet and rill erosion) and for C and P factors on major cropland uses are shown in Table 3-1. In each case, individual values for each factor were extracted from the NRI sample point files in respective cropland uses. The acreage associated with each factor was determined by applying the relevant NRI acreage expansion factor to each sample point.

In Table 3-1, the column labeled RKLS indicates the sheet and rill erosion rates that would be expected if land in each use were continually tilled up and down the slope and left barren of vegetation, a worst-case scenario. On all cropland, average soil displacement under these conditions would be 21.8 tons/acre·year. Cropland used for close-grown crops such as wheat would erode at the lower rate of 15.4 tons/acre·year. This rate reflects the fact that a substantial portion of the close-grown crop acreage is in the Great Plains, where rainfall and R-factor values tend to be lower. Cropland used for hay generally has a high inherent potential for sheet and rill erosion (35.2 tons/acre·year). Use of that land in hay crops, however, alleviates most of the problem, because of the dense soil cover maintained throughout the year, characterized by the low C-factor values associated with sod-based crops. The USLE was developed for a single crop per year; therefore, its application to double-cropping is limited. The distribution of acreage by the RKLS product presents a pattern similar to the distribution of acreage according to sheet and rill erosion rates. The practical significance of

TABLE 3-1 Impact of C and P Factors in USLE Estimates for Sheet and Rill Erosion on Cropland in the United States[a]

Crops	Inherent Erosion Potential, RKLS (tons/acre·year)	Actual Erosion, RKLSCP (tons/acre·year)	Reduction Factor	Percent Reduction	Factors C	P
Row crops	22.3	6.1	3.7	72.6	0.28	0.86
Close-grown crops	15.4	3.2	4.8	79.2	0.20	0.87
Hay	35.2	0.6	58.7	98.2	0.04	0.81
Other	24.5	2.7	9.1	88.9	0.19	0.88
All cropland	21.8	4.3	5.1	80.2	0.26	0.91

[a]Data in each column are average values of many observations, including the values for C and P. Thus, the reductions shown cannot be obtained by multiplying values in each row.

SOURCE: 1982 NRI; adapted from Rosenberry and English, 1986.

this distribution of potential erosion problems and related policy implications are discussed in Chapter 5.

Across the various land uses, including cropland, pastureland, and forestland, the highest average value for the C factor, 0.28 (see Table 3-1), is for land used for row crops. This value reflects the fact that row-crop farming usually involves an annual disturbance of the soil surface in the course of preparing a seed bed. Disturbed land is often exposed for at least a few months to the erosive effects of rainfall. The C-factor value of 0.04 reported for hayland is about one-eighth the average value reported for row crops. The difference explains why average sheet and rill erosion rates on hayland are much lower than on land used for row crops, even though land planted to hay is estimated to be about 40 percent more erodible on average than cultivated cropland.

Average values for the P factor tend to vary only slightly among cropland uses. Average P-factor values are concentrated at the high end of the theoretical range of values for this factor, averaging 0.91 for all cropland. This distribution of P values indicates that a fairly small percentage of cropland acreage has been treated with supporting conservation practices such as contour farming and stripcropping.

Table 3-1 provides an indication of the impact on potential erosion rates of cropping, management, and conservation practices represented by the C and P factors. The column labeled reduction factor is a ratio of the inherent potential erosion (RKLS-product value) and the USLE estimate (RKLS × CP) for each major cropland use. For all cropland, the inherent potential for sheet and rill erosion was 21.8 tons/acre·year, but the effect of the C- and P-factor values was to reduce the estimated USLE rate to 4.3 tons/acre·year. The resulting reduction-factor values in Table 3-1 show that potential erosion rates were reduced most significantly on cropland used for hay (from 35.2 to 0.6 tons/acre·year, a 58.7 reduction factor and a 98 percent decrease in erosion), followed by other land (used mostly for vegetables, orchards, and other crops) and by land used for close-grown crops. Potential erosion was reduced considerably on land used for row crops—by a factor of 3.7 (or a 73 percent reduction)—from an average annual rate of 22.3 to 6.1 tons/acre. The reduction of potential erosion on row-cropped land was less than that for other cropland uses.

The USLE is routinely used by USDA conservation agencies for on-farm planning, program evaluation, and analysis. The committee believes, however, that USDA state- and national-level program management and analytic activities could be strengthened by similarly encouraging routine use of the concept of inherent potential for sheet and rill erosion.

THE MEASURES OF SOIL EROSION

Recently, the SCS has begun to use the concept of inherent erosion potential, as estimated by the RKLS product, in evaluating alternative land classification systems (see Chapter 5). The committee believes that this is a positive step. Promotion of proper use of this concept would advance understanding about erosion problems among soil scientists, conservationists, analysts, and farmers.

The USDA should prepare and encourage adherence to a special publication that presents common concepts, terminology, and definitions of key land use and conservation measures and indicators. Such a publication would thoroughly explain the uses of the concept of *inherent erosion potential* for sheet and rill erosion. Similar guidelines could be issued for wind erosion, when appropriate. The publication could be used by relevant agencies, university researchers, extension service personnel, and other analysts.

The SCS is the appropriate lead agency to draft these guidelines—with the assistance of soil scientists and engineers working for the Agricultural Research Service (ARS), state experiment stations, and other scientific centers. The publication should be distributed to relevant personnel in the ARS, the Economic Research Service, the Forest Service, the Agricultural Stabilization and Conservation Service, and other federal and state agencies.

Key Role of C Factors in Controlling Erosion

Because values for the C factor are contained in the NRI data files for most sample points, the inventory is useful for analyzing the geographic distribution of cropping and management practices and their potential effectiveness in controlling sheet and rill erosion. Analysis of NRI data may also help define further research that is needed on the C factor and help refine factor values.

Recorded NRI values for the C factor reflect probable sheet and rill erosion conditions nationwide and within such smaller geographic regions as Major Land Resource Areas (MLRAs). Ideally, as the potential for erosion (indicated by the RKLS product) increases on cropland, C values should decline, indicating a greater effort at erosion control through cropping and management practices. However, Figure 3-1 demonstrates that average nationwide C values for land with low erosion potential do not differ markedly from values for land with high erosion potential. It also shows the relationship between C- and P-

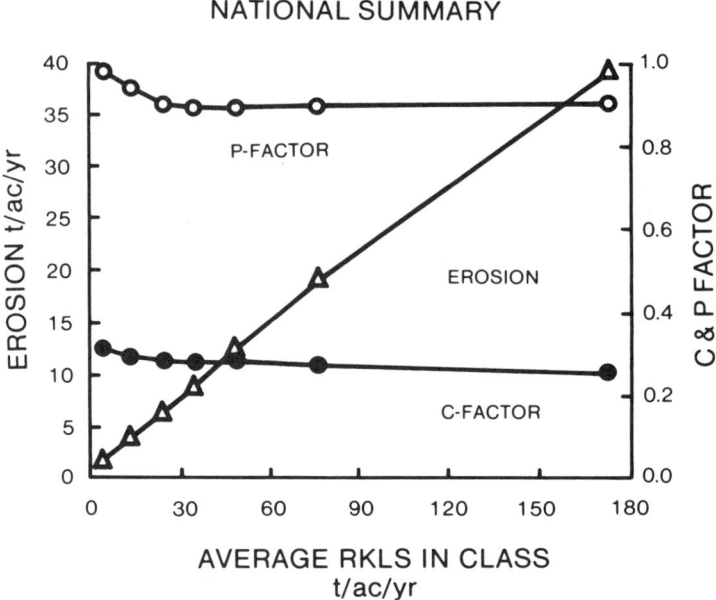

FIGURE 3-1 Plots of the 1982 NRI weighted average erosion rate (tons/acre·year), C and P factor versus the potential for erosion (tons/acre·year) as expressed by the RKLS product of the USLE. Data were summarized nationally from the 1982 NRI. Each data point is plotted at the midpoint of a range or class of RKLS values. Source: Pierce et al., 1986.

factor values and erosion rates as a function of progressively rising inherent erosion potential.

The relationship between C-factor values and inherent erodibility differs across the country (see Figure 3-2). In MLRA 105 (the northern Mississippi Valley loess hills of Wisconsin, Iowa, Minnesota, and Illinois), for example, the average C-factor value is lower than that for the nation as a whole. C-factor values decrease significantly as the potential for erosion increases. The fact that C values change little with inherent erosion potential in MLRAs 103 (the central till prairies of Minnesota and Iowa), 134 (the southern Mississippi Valley silty uplands of Mississippi, Tennessee, and Kentucky), and 136 (the southern Piedmont of Virginia, North Carolina, South Carolina, Georgia, and Alabama), however, indicates that conservation management techniques are not widely used on erosion-prone soils, nor are they concentrated on the most erodible soils.

Similar analyses can be performed for virtually any grouping of cropland. In Figure 3-3 the national average values for the C factor are arrayed by potential erosion rate for class I and subclasses IIe, IIIe, and IVe of the USDA Land Capability Class System (LCCS). (The letter e following the class denotes a subclass of land that has suffered erosion damage in the past or is vulnerable to it.) Under this system, the very best land is designated class I. It has few natural limitations for intensive cultivated crop uses. Successive classes have progressively greater limitations for intensive crop production. Classes IV and above are deemed unsuitable for intensive cropping.

For the nation's class I land, C-factor values generally decline as the erosion potential increases. However, on subclasses IIe, IIIe, and IVe, which comprise the majority of erodible U.S. cropland, the C-factor values remain fairly constant as erosion potential increases. This find-

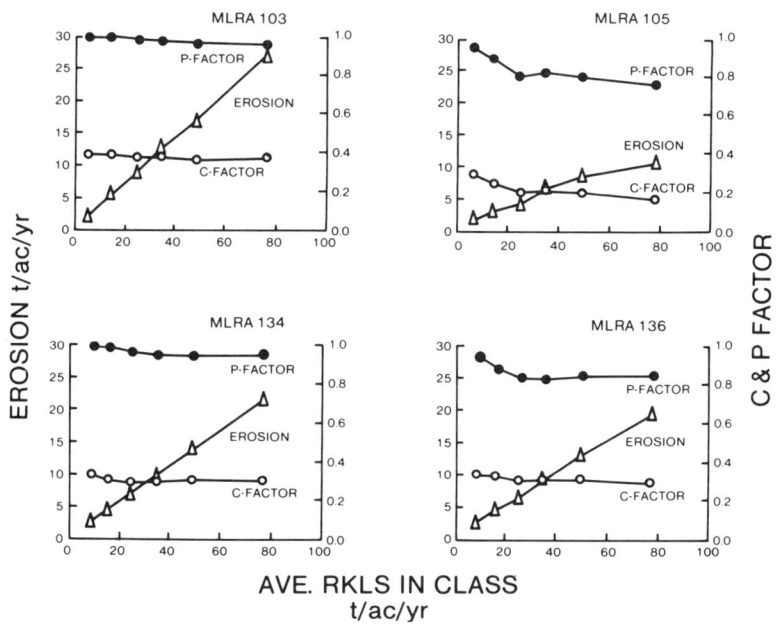

FIGURE 3-2 Plots of the 1982 NRI weighted average erosion rate (tons/acre·year), C and P factor versus the potential for erosion (tons/acre·year) as expressed by the RKLS product of the USLE for MLRAs 103 (central Iowa and Minnesota till prairies), 105 (northern Mississippi Valley loess hills), 134 (southern Mississippi Valley silty uplands), and 136 (southern Piedmont). Source: Pierce et al., 1986.

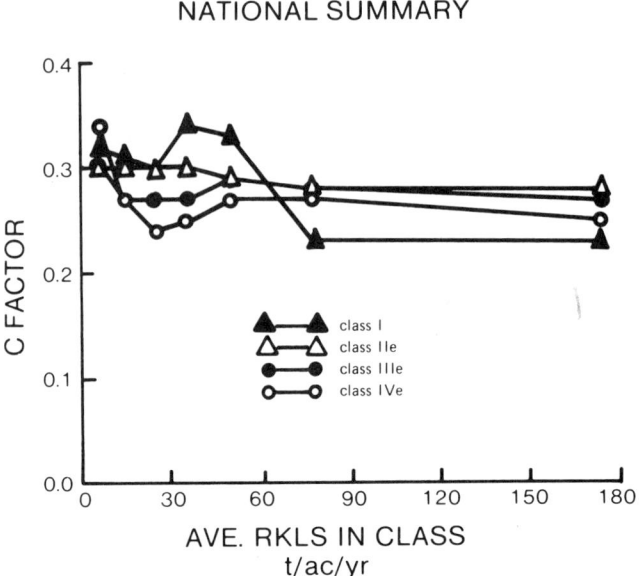

FIGURE 3-3 Plots of the weighted average C factor versus potential for erosion (tons/acre·year) as expressed by the RKLS factor of the USLE for land capability subclasses I, IIe, IIIe, and IVe. Data were summarized nationally from the 1982 NRI. Source: Pierce et al., 1986.

ing suggests that management and cropping practices beneficial in reducing soil erosion have not been concentrated on those lands that are in greatest need of erosion control.

Uncertainties Associated With C-Factor Values

A sizable scientific literature exists on the USLE and on the characteristics and proper values of each of its individual factors. As noted later in this chapter, further research is needed on concentrated flow and on the relationship between erosion and the transport of sediment from farm fields to watercourses.

The equation is continually being refined to reflect specific climatic, soil, or vegetative conditions and improvements in the understanding of the dynamics of water erosion processes. Such work will continue to be necessary, because land use and tillage practices will change over time.

To evaluate the validity and usefulness of the USLE estimates included in the 1982 NRI, the committee thought it advisable to examine some of the conceptual and empirical issues associated with contemporary C-factor values. Values assigned to C factors are critical determinants of sheet and rill erosion estimates in the USLE. Those values thus have important scientific and policy implications. Although a considerable amount of research has been directed toward refining C-factor values and testing C factors in field experiments, the committee believes that they can be improved through further revision and refinement. The research community, including scientists within the ARS and state agricultural experiment stations, must take a leadership role in carrying out this work.

Theoretically, the range of possible variation in C-factor values, from 0.001 for undisturbed forestland to 1.0 for clean-tilled, fallow land, makes it the most important variable controlled by human activity in determining estimated sheet and rill erosion rates. In application, however, the full range of C-factor values is primarily most useful when erosion rates are compared among land uses—for example, when land is converted from row crops (average C value of 0.28) to permanent pasture (C value of 0.01 or lower). Within a given land use, C-factor values for most cropland are within a narrower range. For cropland used for row crops, for example, nearly 57 percent of the acreage has C-factor values between 0.25 and 0.45 (Rosenberry and English, 1986). The C-factor values for a particular subcategory of land use—cropland used to grow corn, for example—would tend to be even more tightly clustered around the mean value.

Despite the narrower range for C-factor values for a given crop or land use, it is important to emphasize that of all the factors in the USLE, the C factor is the one most subject to change as a result of changes in farming practices. Continuous corn production with conventional tillage might yield a C-factor value of 0.37, for example; theoretically, corn produced on comparable land in the same area under no-till planting in sod would be assigned a C value 37 times lower (C = 0.01). If sheet and rill erosion were 50 tons/acre·year under the conventional system, it would be reduced to less than 2 tons/acre·year through use of the no-till method.

Mulch-Factor Value The C-factor values encountered in the NRI data bases are subject to several types of errors. One source of bias arises from the relationship between the percentage of residue cover on the

Corn growing in soybean residue in a no-till system, Jackson County, Iowa. No-till C-factor values are greatly reduced compared with those for continuous corn production with conventional tillage. Credit: U.S. Department of Agriculture, Soil Conservation Service.

soil surface and its corresponding mulch, or residue, value. The mulch-factor value accounts for the erosion reduction produced by crop residues buried just below the surface. As contained in the USLE, the relationship tends to skew estimated rates of erosion upward by overestimating C-factor values under some conditions. The mulch value is used to adjust C-factor values in the equation to reflect differences in tillage systems.

Recent research (Pierce et al., 1986) indicates that, for a given percentage of residue cover on the soil surface, the mulch factor can be quite variable (ranging from 0.5 to 0.01), depending on the roughness of a soil surface and the type of residue and extent of its incorporation into the soil. Others have shown less variation. The more extreme findings indicate that values for the mulch factor might be three to four times too high in the USLE used in the NRIs. It implies that the values of the C factor—and, hence, estimated erosion rates—for cropland reported in the NRI are also too high. The bias is most pronounced when the percentage of surface cover is relatively low (from about 15 to 30 percent). Thus, where fairly high C-factor values are encountered in the NRIs, sheet and rill erosion rates might be overstated.

Field Application of the USLE Other sources of error in the values for the C factor arise as a result of the application of the equation in the field. Ascribing the correct value for C for a particular location is largely subjective. A standard method is to measure the amount of ground cover along a transect. However, a recent study of variability in residue cover revealed considerable variation among transects for the same field—and among observers along the same transect (Richards et al., 1984). A study of the application of the L, S, and K factors would probably reveal similar variations on fields that are difficult to characterize because of variable soils and topography.

In the 1982 NRI, as in most field applications of the USLE, SCS personnel generally did not attempt to measure the amount of ground cover. Rather, they determined the type of tillage practice used at a specific sample point and then recorded the appropriate C-factor values from standard generalized tables. Field studies have shown, however, that these tables may be misleading. The use of specific tillage implements or practices can result in a wide range of crop residue levels on the soil surface. A review (Colvin et al., 1981) of three tillage studies found that spring tillage with a disk harrow resulted in crop residue cover ranging from 42 to 73 percent of the soil surface. Such widely ranging crop residue levels have very different implications for soil erosion conditions and raise questions about the actual erosion control afforded by specific practices.

Conservation tillage is a case in point (see Chapter 1 for definitions of tillage systems). Estimates of acreage treated with conservation tillage in the 1982 NRI (and with minimum tillage in the 1977 NRI) are based primarily on assignment of C-factor values for that practice taken from standard tables. Had actual residue levels been measured for most sample points, very different C-factor values and resultant erosion rates might have been recorded.

A 1980 survey by Peter Nowak (University of Wisconsin, Madison, personal communication) indicates that erosion control benefits might be overstated when field interpretations of this kind are used (see Pierce et al., 1986). Nowak interviewed 200 farmers from three Iowa watersheds; 78 percent professed to use conservation tillage systems.

Upon checking residue levels on their farms, however, Nowak found that only 7 percent of the corn acreage and 26 percent of the soybean acreage met SCS technical residue specifications for conservation tillage. If in fact what is reported as conservation tillage is not achieving expected levels of erosion control implied by the term, such tillage might be providing considerably less erosion control on less acreage across the United States than is suggested by the rapid increase in the use of conservation tillage equipment (chisel plows, for example, and implements other than moldboard plows).

Errors of Omission Finally, C-factor values are subject to what might be called errors of omission. In some parts of the United States, existing C-factor values do not adequately represent vegetation-erosion relationships. In certain rangeland areas, for example, a pavement effect or flat rock fragments on exposed soils actually protect those soils from the erosive forces of raindrops and runoff, though nearby soils can be eroded as runoff concentrates along natural channels and drainageways. Some researchers believe that this phenomenon should be reflected in K rather than C factors. Others propose that a subfactor should be applied to the USLE in rangeland watersheds prone to soil pavement and concentrated flow effects. The subfactor approach also has been proposed to improve the USLE procedure and C-factor values for forest conditions in the southeastern United States.

Many uses can be made of this information for scientific and policy analysis and program design, administration, and evaluation. Rates and total tons of soil displaced through sheet and rill erosion can be tabulated for any land use, crop, or land capability class and can be aggregated for the entire country, states, or MLRAs. Acreage in a given land use can be tabulated according to its erosion rate or its erosion rate in relation to its T value. Erosion rates under alternative cropping and conservation conditions can be simulated for any land use (see the boxed article Erosion Under Alternative Cropping, Management, and Conservation Practices).

Recommendations on the USLE

The data in the 1977 and 1982 NRIs can be put to many uses, including those illustrated in this report. The sheet and rill erosion information contained in the 1982 NRI is reliable and sufficiently accurate for many analytical applications at the national, regional, and state levels. Those uses include conservation program planning and analysis, refinement

Erosion Under Alternative Cropping, Management, and Conservation Practices

The sheet and rill erosion estimates in the NRIs reflect values for the C factor recorded at each sample point. By altering those C-factor values in the computerized NRI data base, it is possible to simulate erosion rates that would be expected from alternative cropping and management conditions or the addition of supporting conservation practices. In effect, the USLE files for any geographic aggregation in the NRIs can be used to ask questions about the alternative effects of varying combinations of these conditions. For comparative purposes other factors such as slope, length, and soil materials are constant.

A sample analysis is presented in Figure 3-4. The C-factor values for all NRI sample points falling on land in row and close-grown crops were altered in the NRI data files to reflect a range of C-factor values. At one end of the continuum, with uniformly assigned C-factor values of 0.30 (the national average value for row and close-grown crops), about 73 percent of this acreage would have sheet and rill erosion

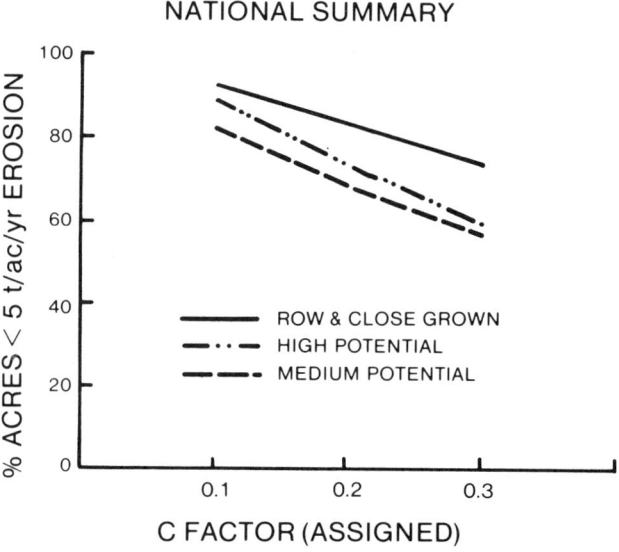

FIGURE 3-4 Percentage of acres nationally with USLE erosion rates greater than 5 tons/acre·year at assumed levels of C factors for land in row and close-grown crops in 1982. Note: High- and medium-potential curves illustrate land with a high or medium potential for conversion to cropland. Source: Pierce et al., 1986.

rates of less than 5 tons/acre·year. This corresponds closely to the estimated 75 percent of the row and close-grown crop acreage reported as eroding below 5 tons/acre·year in the 1982 NRI.

Under a more optimistic assumption, all land in row and close-grown crops is assigned a C-factor value of 0.1, about the average value expected if all land were farmed by no-till methods—the ultimate form of conservation tillage—and if heavy crop residue levels were maintained throughout the cropping year. Under these circumstances about 93 percent of the 323 million acres would experience sheet and rill erosion rates well below 5 tons/acre·year. At least 7 percent of the land now in principal crops—over 20 million acres—would still erode at rates above 5 tons/acre·year, even if they were farmed with the most effective, commonly available technology. Some additional structural practices including stripcropping, terracing, or permanent vegetative cover would be required to bring erosion rates to within the 5-ton maximum T-value level on these lands.

Similar analyses at the MLRA level reveal striking differences in the erosion rates under alternative farming conditions. In such heavily cropped parts of the country as MLRA 103 (the central Iowa and Minnesota till prairies) a much higher proportion of land would be adequately protected from erosion with an average level of conservation management (see Figure 3-5). By contrast, in MLRA 105 (the northern Mississippi Valley loess hills of Wisconsin, Iowa, Minnesota, and Illinois) and MLRA 136 (the southern Piedmont areas of Virginia, North Carolina, and South Carolina, Georgia, and Alabama), even a high degree of conservation management would leave 20 to 40 percent of the acreage vulnerable to erosion rates in excess of 5 tons/acre·year.

Similar analyses can be performed to simulate the effect of converting land from use for row and close-grown crops to less intensive uses. As part of the 1977 and 1982 NRIs, officials of the local SCS, the conservation district, and the extension service assessed the potential for future conversion of pastureland, forestland, and rangeland to cropland use, based on the physical characteristics of the land and recent patterns of land use in each county or conservation district. Because USLE data were recorded for this potential cropland, it is possible to estimate what sheet and rill erosion conditions would be like if the land were converted from its present use to more intensive crop uses in the future.

Results of such an analysis for land with high and medium potential for conversion are shown in Figures 3-4 and 3-5. Even the high-potential cropland presents, on average, a more significant erosion control

THE MEASURES OF SOIL EROSION

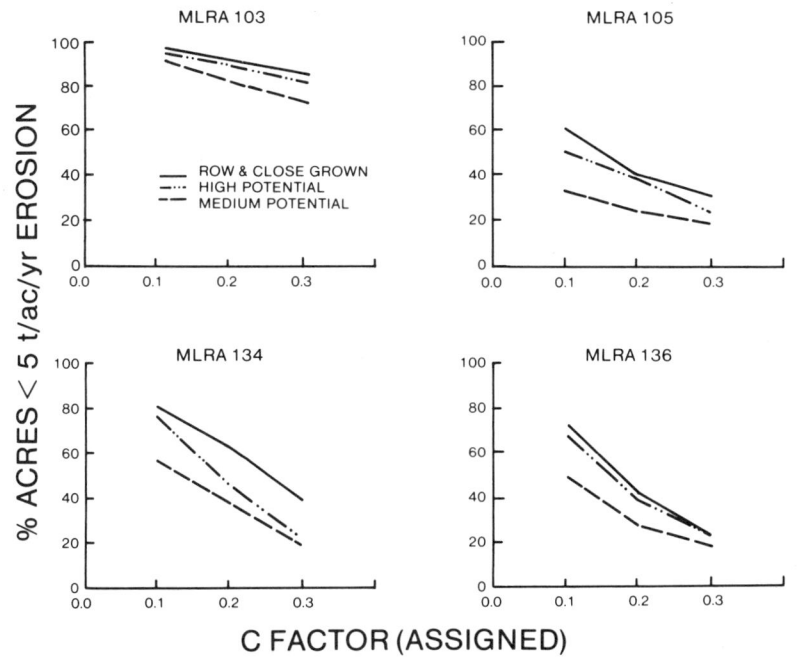

FIGURE 3-5 Percentage of acres in MLRAs 103 (central Iowa and Minnesota till prairies), 105 (northern Mississippi Valley loess hills), 134 (southern Mississippi Valley silty uplands), and 136 (southern Piedmont) with USLE erosion rates greater than 5 tons/acre·year at assumed levels of C factors for land in row and close-grown crops in 1982. Note: High- and medium-potential curves illustrate land with high or medium potential for conversion to cropland. Source: Pierce et al., 1986.

challenge than land used for row and close-grown crops in 1982. With a C-factor value of 0.30—the average conditions for land in row and close-grown crops in 1982—only 60 percent of the country's high-potential cropland and 57 percent of the medium-potential cropland would have sheet and rill erosion rates less than 5 tons/acre·year. Even with much lower C-factor values of 0.1, representative of no-till farming systems, 11 percent of the high-potential cropland and 18 percent of the medium potential cropland would be expected to have erosion rates greater than 5 tons/acre·year. Among regions, considerable variation is found in the vulnerability of high- and medium-potential cropland to average annual erosion rates greater than 5 tons/acre.

> These analyses of alternative C-factor values indicate the diversity of sheet and rill erosion problems across the country. No single cropping and management approach, including no-till farming, will solve the erosion problem on all cropland. These analyses also suggest the value of the 1977 and 1982 NRIs for assessing soil erosion conditions and alternative conservation practices. Similar analyses can be performed to assess the erosion conditions that would be expected under alternative conservation supporting practical conditions, such as stripcropping, by altering P-factor values in the NRI data files.

of procedures for estimating erosion rates, and analysis of basic erosion processes and effects.

The USLE data can also be used to test new techniques that might be incorporated in future resource inventories to estimate erosion or assess other phenomena, such as nonpoint pollution. While erosion estimates based on USLE data in the 1982 NRI are site specific and indicate soil displacement rather than direct erosion damages, the estimates are sufficiently accurate to serve as primary guides for state, regional, and national analyses of conservation needs and opportunities until better analytical tools and data become available.

Nevertheless, the committee recommends that efforts continue to improve the USLE, for conservation planning and future resource inventories. A special need exists to improve the accuracy of the equation, or to develop a separate equation, for areas where water sources other than rainfall, including snowmelt and irrigation, are important contributors to total soil displacement. Further work—much of it under way—is needed to improve the methods of measurement and accuracy of individual factors in the equation. In addition, as noted earlier, the relationship between sheet and rill erosion, and concentrated flow remains uncertain.

Some of the limitations of USLE data addressed in this report stem from efforts over the last decade to apply the USLE in ways for which it was not originally designed. The equation was not designed to directly measure damages from erosion, either onsite or offsite. The committee notes that several mathematical models of erosion-productivity interactions (the Productivity Index model and the Erosion Productivity Impact Calculator) have been designed to make use of NRI data. Similarly, models such as the Agricultural Runoff Model (ARM); the Hydrologic Simulation Program in Fortran (HSPF); the Nonpoint Source Model (NPS); and later, Chemicals, Runoff, and Erosion from Agricultural Management Systems (CREAMS) have been developed, which incorporate USLE data for evaluation of offsite erosion effects using NRI data. Other models need to be assessed; new models should be developed, building on and extending USLE data for specific applications.

The SCS and other agencies at the USDA should continue to rely on the USLE as the major analytical device for conservation program planning, management, and analysis. SCS should take the lead in continuously upgrading field-level expertise in USLE applications. The agency should also introduce new techniques as soon as they are available that more directly relate erosion rates observed in the field to actual onsite and offsite damages.

The committee believes that the erosion equations and NRIs have not yet been used to full advantage. If continued public investments in these tools are to be justified, the committee believes that the USDA has a continuing obligation to demonstrate that important, practical benefits are accruing from the use of these tools in ongoing efforts to control serious erosion problems.

Sheet and rill erosion problems on much of the land currently used for row and close-grown crops often can be adequately controlled through changes in cropping and management practices. Conservation tillage—no-till farming systems, in particular—demonstrates exceptional promise for erosion control, provided that adequate levels of crop residues are maintained.

On many soils, additional soil conservation benefits can be achieved by using traditional supporting conservation practices such as contour farming systems, vegetated waterways, terracing, and stripcropping in conjunction with conservation tillage systems. However, rainfall, type of soil, and slope characteristics beyond the control of farmers make it extremely difficult to economically control erosion on the approximately 30 million acres of erosion-prone soils currently in conventional cultivation. The committee stresses that basic land use patterns are of great significance in conserving highly erodible lands. Where the potential for sheet and rill erosion is great, economically feasible control of erosion levels can be achieved only through shifts away from conventionally cultivated crops and into such uses as pasture, hay, range, some tree crops, or forest.

Wind Erosion Estimates

The 1977 NRI provided estimates of wind erosion for the 10 Great Plains states, using the WEE described by Skidmore and Woodruff (1968). In the 1982 NRI, wind erosion estimates were attempted for all nonfederal lands in the United States on all land uses (see the boxed article The Wind Erosion Equation).

The Wind Erosion Equation

The version of the WEE used in the 1982 NRI is a variation of equations developed in the early 1960s. It has the following functional form:

$$E = f(I, C, K, L, V)$$

Unlike the USLE, the factors in the WEE are not directly multiplied together to determine a value for E, which is the potential annual wind erosion rate in tons/acre year. Rather, this potential is a function of the following:

I The soil erodibility value, or I factor, reflects the size of soil particles or aggregates. The value for the I factor is the single most important source of variance in the equation. The I-factor value is expressed as the average annual soil loss expected to occur from an isolated, level, smooth, unsheltered field, barren of vegetation and with a noncrusted surface. Originally, the I factor was derived from the total annual erosion (in tons/acre·year) for a number of farm fields located near Garden City, Kansas. In applications such as the NRI, the I values are assigned to soils based on the percentage of soil mass occurring in aggregates smaller than 0.84 mm in diameter. This value is obtained from SCS technical guides, in which values of I correspond to wind erosion groups (WEGs). The WEGs are in turn based on predominant soil textural classes (percentages of sand, silt, and clay), which were determined by reference to soil surveys or other technical guides.

C C is an adjustment value to correct for areas having wind speed and rainfall patterns different from those of the reference area used to construct the original equation. The correction is based on mean quantities of wind speed and rainfall evaporation. The value for C is obtained from SCS technical guides.

THE MEASURES OF SOIL EROSION

K The soil ridge roughness value, K, is assigned to a sample point based on field inspection. The lower the value, the greater the vulnerability of a field to erosion, if other conditions are equal. Deeply furrowed fields tend to trap wind-blown soil in their furrows. Only two K-factor values could be assigned in the NRI: K = 0.05, for a smooth field, and K = 1.0, for a ridged field.

L The value representing the unsheltered distance across a field along the prevailing wind direction is L. The unsheltered area of a field begins leeward of a protected area or from a barrier at a distance of 10 times the barrier's effective height and perpendicular to the prevailing wind direction. According to the WEE, a field windbreak composed of trees 40 feet in height will theoretically protect a field on the leeward side of the trees for a distance of 400 feet. This value is based on field inspection.

V The vegetative cover value V combines residue quantity, type, and orientation (flat or standing). The correct value is selected from SCS technical guides based upon field inspection.

Sand moves across the highway in 50 mph winds near Pacific City, Oregon. The new beach grass planting (left) will offer some protection against wind erosion. Credit: U.S. Department of Agriculture, Soil Conservation Service.

The development of the WEE marked a major conceptual advance in soil erosion science. Use of the equation in the 1977 and 1982 NRIs was an important step toward identifying lands subject to serious damage from wind erosion. The committee believes that while the wind erosion data contained in the NRIs are useful for a number of important scientific and policy applications, the quantitative estimates are not sufficiently reliable for many uses. For example, the NRI estimates probably provide reliable indicators of the relative hazard from wind erosion in parts of the United States, especially for areas in the Great Plains states where wind erosion is chronically severe. It is likely, however, that the absolute values are too high (Gillette, 1986). They do not appear to provide accurate estimates in more humid regions, on heavier soils, or in some arid regions where conditions of the soil surface in relation to wind erosion are not yet well defined.

Limitations of the WEE

Compared with the USLE data, the 1977 and 1982 NRI data for wind erosion are less accurate and less reliable. Moreover, because of limitations of time and personnel, as well as added computer costs, values for the individual factors of the equation are not currently part of the computerized NRI data bases. For this reason, it is not yet possible to use NRI data directly to refine the factors or the overall equation, or to undertake analyses of alternative options described previously in reference to the USLE.

There are several reasons for the present variation in estimates of sheet and rill erosion rates in contrast to those for wind. First, scientists have identified a number of conceptual shortcomings of the equation used to estimate wind erosion, some of which are related to the relatively narrow empirical and experimental basis of the WEE (Gillette, 1986) (see the boxed article Improving Estimates of Wind Erosion). Second, wind erosion estimation techniques used in the field have received less attention from scientific researchers than techniques for sheet and rill erosion, despite great interest in wind erosion at the inception of the modern soil conservation movement in the 1930s. Until uncertainties are resolved, it will be difficult for analysts to interpret estimates of wind erosion in the 1982 NRI.

Improving Wind Erosion Prediction

An expanded research effort is needed to develop and validate basic components useful in improving the WEE. (Some avenues of research

Improving Estimates of Wind Erosion

To identify some of the problems and research needs in the area of wind erosion prediction, the committee commissioned a paper for the project on the subject of the WEE and its use in the 1982 NRI (Gillette, 1986). A summary of findings follows.

Several sources of possible errors in the WEE, and thus in the NRI data sets for wind erosion, were identified. Some new work has been completed on the threshold wind velocities required to initiate erosion of soil particles of various sizes. This research indicates that the I, or erodibility, values used in the WEE might lead to systematic overestimates of erosion when 20 to 65 percent of the surface soil mass is composed of particles smaller than 0.84 mm in diameter. If this finding is correct, the NRI wind erosion estimates would generally be too high for certain soils, particularly those with textures other than sand and sandy loam. In practical terms, the estimated average annual erosion rates reported for humid areas in the 1982 NRI are generally low. Yet those estimates are still probably too high.

A single value for I is used for an entire year when the WEE is used. However, soil aggregates can alter in one season. During drought, for example, soil aggregates of clay-textured soils from western Texas disintegrated. As the average size of aggregates decreased, the threshold wind velocity decreased, too.

The C factor in the WEE is used to adjust the I factor to reflect climatic circumstances of wind speed and rainfall evaporation for areas other than those found in the reference area (Garden City, Kansas). Recent research on wind threshold velocities indicates that the C factor leads to overestimates of wind erosion wherever mean wind speeds are lower than those for Garden City. Wind speeds are in fact lower for most regions of the United States.

An attempt was made to compare wind erosion estimates for a humid area (Minneapolis) using the WEE with estimates derived from a provisional WEE that reflects new research on wind threshold velocities. The results showed moderately good agreement for soils in that area. The supplemented WEE predicts that the soils generally susceptible to high rates of wind erosion are very fine-to-medium sands or sandy loams in WEGs 1 and 2. Such soils do not predominate in this midwestern area. The WEE appears to overstate wind erosion by a factor of about five for the more common moderately erodible soils in WEGs 3, 4, and 4L (textures ranging from very fine, sandy loams to silty clay loams). For soils in WEGs 5 and 6, the WEE appears to overestimate wind erosion rates by a factor of about 10.

have been suggested by Gillette [1986].) In the near term, however, efforts should be made to compare application of the WEE in the NRI with other methods. Comparisons might help identify the need or the opportunity to adjust NRI wind erosion estimates to make them more accurate and usable.

Values for individual factors of the WEE should be coded and made available on a special supplemental NRI computer tape, if at all possible. They should also be available on the raw data tapes during any future inventories, at least for some regions.

As a first step, wind erosion factor data should be entered for selected MLRAs representative of a diverse range of wind erosion conditions. Such data would facilitate basic research on wind erosion prediction and lend greater authority to short-term policy and program applications in arid regions. These computer tapes should be made available to interested analysts inside and outside of the USDA.

The USDA should design and fund a comprehensive plan of research to improve the prediction of wind erosion over the long term and to compare NRI wind erosion estimates with those derived by other methods during the next few years. A similar program should be developed to improve water erosion prediction.

The research plan should specify the roles and responsibilities of the SCS, the ARS, experiment stations, university scientists, and scientists from other agencies. A realistic mechanism for coordination, along with solid USDA commitment and leadership, will be essential to ensure that the research program is effective, efficient, and of high quality. The overall research plan and strategy for evaluating and improving wind erosion estimation techniques ideally should be developed—or at least reviewed—by a panel of experts working outside of USDA. Field verification should be performed by independent contractors, using criteria developed by an independent panel of experts.

The committee recommends that this research program include

- An assessment of an alternative formulation of a WEE;
- Verification of WEE estimates, basic parameters, and concepts through field-level measurements;
- Study of the deposition patterns of soil eroded by wind, with a special focus on air and water pollution consequences;
- Reconsideration of the need for nationwide wind erosion data.

It might be possible to reliably apply certain variables of the WEE, perhaps with modifications, as indicators of relative wind erosion hazard. Such an approach has been proposed by the Resource Conservation Act (RCA) Fragile Soils Work Group as a means of identifying highly erodible lands (McCormack and Heimlich, 1985). This group of USDA soil scientists and analysts evaluated six major options for classifying erodible soils. The recommended option uses the soil loss tolerance (T) factor and soil and site factors of the USLE and the WEE as the basis for defining erodible soils, RKLS/T and CI/T, for example.

Some method for classifying land according to its relative wind erosion hazard will be needed if government agricultural policies are revised in the future to take soil erosion problems into account. The NRI could provide a useful data base for evaluating some of the effects of sodbuster and conservation reserve initiatives, for example, in areas where wind erosion is a problem. First, however, some scientific uncertainties about the WEE and NRI wind erosion estimates must be resolved.

Evaluation of the usefulness of the WEE and its individual factors as criteria for identifying lands that are highly erodible by wind should be a priority of the USDA. The evaluation should build upon the work of USDA's RCA Fragile Soils Work Group. Sufficient personnel and resources should be allotted to this activity to enable the USDA to propose in 1986 scientifically defensible and administratively feasible criteria to identify highly erodible land that is currently in cultivation or subject to cultivation in arid or semiarid regions. The rapid development of a representative computer data base of WEE-factor values collected in the 1982 NRI would expedite and support this activity.

Erosion by Concentrated Flow: Ephemeral Gullies

Soil scientists, conservationists, and farmers have recognized for many years that serious erosion problems sometimes occur within natural drainageways, usually on cropland where water runoff concentrates and washes soil from wide, shallow channels. Unlike smaller rills, the sites of these concentrated flow channels are relatively permanent topographic features. Like rills, however, tillage operations usually obliterate evidence of concentrated flow erosion; the term *ephemeral gully erosion* reflects this condition.

Ephemeral gully areas tend to erode, sometimes at very high rates, unless appropriate steps are taken to divert water runoff or to plant the channel area in a sod-type crop. Establishing grass-based sod in these areas is a highly effective erosion control practice, and they are referred to as grassed waterways.

There is no commonly accepted, practical method for estimating ephemeral gully erosion in the field; USLE estimates of sheet and rill erosion do not independently predict or quantify ephemeral gully erosion. Neither the 1977 nor the 1982 NRI includes estimates of ephemeral gully erosion. The extent and magnitude of this type of erosion are not well established, because methods for doing so are lacking.

In some areas, however, this form of erosion is thought to equal or surpass others as a soil conservation problem. Evidence from field observations, measurements, and water quality models suggests that ephemeral gully erosion can result in sediment removal rates comparable to those for sheet and rill erosion (Foster, 1986). (Sheet and rill erosion can be soil displacement—not necessarily soil loss—from a given field.) Concentrated flow may accentuate sheet and rill erosion in adjacent areas and add to crop loss within the gullied area. The flow itself may transport sediment and chemicals directly from fields to water courses.

In some cases, conservation tillage is adequate to control ephemeral gully erosion. More serious problems may require grassed waterways to directly protect the eroding channel or the construction of terraces, diversions, and outlets to slow and divert runoff as it moves downslope. Maintenance of grassed waterways, however, can be a problem in farming systems that use herbicides that are toxic to grasses. Special structures—in effect, small dams and spillways—are sometimes installed to prevent some ephemeral gully areas from evolving into deep, permanent gullies.

SCS personnel in about 30 states are making field estimates of ephemeral gully erosion. Several estimation methods are being used (Foster, 1986). Plans are being developed to assemble all data for analysis within the coming year.

In the committee's judgment, there is reason to believe that ephemeral gully erosion may pose significant problems on cultivated cropland of sloping, uneven topography. Ephemeral gully erosion may prove to be the predominant and, perhaps, most damaging form of erosion in some regions, in terms of soil productivity and offsite problems.

A comprehensive and coordinated research program involving the ARS, the SCS, experiment stations, the Extension Service, other federal agencies, and universities should be developed to meet the following objectives:

• Adoption of a consistent terminology to describe ephemeral gully erosion that will help distinguish it from other forms of erosion.
• Evaluation of field methods for estimating ephemeral gully erosion. Initially, evaluations should emphasize the means for identifying areas affected to varying degrees by ephemeral gully erosion rather than procedures for estimating rates of erosion with great precision. In

THE MEASURES OF SOIL EROSION

many cases, identification of specific problems, combined with general (as opposed to site-specific) information about the damages associated with those problems, will aid in developing recommendations for land users regarding the adoption of cost-effective conservation measures.

- Acceleration of research and educational programs conducted by the ARS, the SCS, experiment stations, universities, the extension service, and other USDA agencies on ephemeral gully erosion in regions where this problem is widespread.

- Assessment of deposition patterns associated with ephemeral gully erosion to ascertain the extent of any changes in the productivity of landscapes where soil sediment is deposited.

- Development of an interagency work group should be convened within USDA to study the feasibility of including methods for estimating the extent and severity of ephemeral gully erosion in future NRIs. It may be desirable to focus future inventory activities on areas where preliminary assessments identify serious ephemeral gully erosion problems. One criterion for severity should include the potential impact of ephemeral gully erosion on water quality.

To the extent that areas and specific sites can be identified where ephemeral gully erosion constitutes a significant problem:

- Greater emphasis should be placed on targeting cost-sharing programs and technical assistance to conservation systems that include grassed waterways and other proven, affordable methods for controlling ephemeral gully erosion.

4
On-Farm and Off-Farm Consequences of Soil Erosion

After several decades of relative quiescence, many new, innovative research studies on soil erosion control have been initiated. Some studies (Larson et al., 1983; Williams et al., 1981) have focused on the effects of erosion on soil productivity. Less research is currently under way on offsite damages associated with soil erosion, which was a major area of interest in the 1930s and 1940s. Offsite damages are now an issue that some suggest will likely represent the most serious social consequence of soil erosion on much of cultivated U.S. cropland (Crosson, 1984; Clark et al., 1985). The committee believes that a strong case can be made to initiate more work on offsite damages. New research might include assessments of the chemical composition of runoff and the effects of chemical constituents and sediment derived from erosion on water quality and the ecology of rivers, streams, and reservoirs.

General findings of recent research on the major on-farm and off-farm consequences of soil erosion, focusing on research applications of NRI data, are summarized in this chapter. This research—much of which was possible because of the availability of NRI data—attempts to quantify basic reasons for conserving soil: reducing the impact of soil erosion on farm production costs and productivity in terms of yield and mitigating costs associated with offsite damages caused by erosion.

The committee believes that both on-farm and off-farm costs of erosion are important. In some areas, erosion causes onsite productivity losses and also contributes to nonpoint pollution problems. In other areas, erosion might cause only one of these problems. These distinctions are critical to the design of effective conservation policies.

Effects of Erosion on Production Costs

Short-Term Costs

Soil erosion and associated water runoff increase short-term farm production costs per unit of harvested crop in a variety of ways. Water runoff and sediment loss frequently displace fertilizer nutrients and pesticides from the area of original application. Yields may be reduced in these areas as a result of nutrient deficiencies, lack of sufficient moisture, or weed and insect problems. Farmers might be able to correct some or all of these problems, but not without incurring additional production costs.

The severity of these problems is rarely uniform within individual farm fields. It is often impractical for farmers to identify and efficiently correct soil erosion and runoff problems, because they usually follow a common management routine on a given field. Problems on portions of fields are generally tolerated because of the prohibitive cost of more precise management practices. In most cases, to correct runoff problems farmers have three options: (1) use conservation practices to bring erosion under control; (2) change land use, crop rotation, or both; or (3) alter field boundaries. The last two options are often not economically feasible considering current world market prices.

Erosion can also directly damage crops, especially newly planted crops. The damage might be confined to an area of concentrated flow, where seedlings are washed away or inundated with sediment. Or it might affect substantial areas. This is often the case in the Great Plains where wind-blown soil particles bury or abrade germinating crops. In areas of rapidly developing severe gully problems, farm production costs can be increased through damages to farm equipment from crossing gullies and the greater fuel and labor requirements needed to farm around gullies.

Except in areas of concentrated flows or where wind abrasion is severe, short-term effects of erosion on farm production costs and profits can be gradual and subtle. Researchers and farmers can find it difficult to accurately diagnose the effects of erosion on productivity or to distinguish them from other positive and negative influences such as weather and changing technology. Conservationists can often characterize erosion-induced problems in physical terms, but they cannot provide reliable estimates of the short-term economic costs associated with them.

The effects of erosion occur over time; thus, short-term costs are modest on most cropland. The committee believes, however, that bet-

Severe erosion from the concentrated flow of water appears on this unprotected cropland after spring rains (Montgomery County, Iowa). Soil erosion and associated water runoff can increase short-term farm production costs in several ways, including loss of plant nutrients and damage to machinery incurred when crossing the channels left by runoff. Credit: U.S. Department of Agriculture, Soil Conservation Service.

ter methods are needed to estimate these costs on cropland that is subject to high erosion rates. Improved methods would help farmers integrate the full economic costs and benefits of conservation practices into their economic planning.

Long-Term Costs

Long-term increases in the cost of farm production occur when erosion severely, and sometimes permanently, alters the productive capacity of a soil to support physical and economic crop production. The adverse effects of erosion on the depth and nature of the rooting zone available to plants are probably the most pervasive long-term cause of soil productivity losses. The most serious impacts occur when erosion reduces the depth of already shallow topsoils underlain by inhospitable clay subsoils or other material unfavorable to plant growth, or reduces shallow soils underlain by bedrock.

Long-term consequences are often subtle. Mixing of subsoil that gradually becomes a more important part of the rooting zone—a result of progressive erosion—rarely makes the soil totally inhospitable to plants. Such eroded sites usually have less capacity to supply soil moisture or plant nutrients in forms readily available to growing crops. With time, farm production costs increase because more fertilizer, more careful tillage operations, or other changes in management are required to attain a given crop yield. The most serious and ubiquitous long-term consequence of erosion, however, appears to be a reduction in the amount of water in the root zone and increased susceptibility to drought.

The forces of wind and water can alter the physical properties of surface soils themselves. Organic matter in the surface soil is selectively eroded by wind and water. As subsoil material low in organic matter is mixed with topsoil by tillage, the organic matter content of the topsoil is diluted. The common result is a reduced capacity of surface soils to retain nutrients and moisture. The impact of raindrops on bare soil, especially if it is low in organic matter, can cause the formation of a crust or pavement. This and other structural changes in the soil can impede infiltration of rainfall, causing increased runoff and accelerated erosion.

As is the case with short-term erosion damages, long-term damages may be confined to fairly small portions of a given field, which often are easy to spot because of physical changes such as soil color and chronically lower yields. Rather extensive potential damage, however, can go unnoticed. Changes in agricultural technology and management practices confound recognition and quantification of long-term costs of erosion. In fact, crop yields have increased substantially over time in many areas where erosion rates are high. This paradox has led some observers to discount the importance of on-farm productivity losses from erosion. Extended periods of surplus crop production have reinforced this conclusion.

Erosion probably exerts minimal adverse impacts on productivity in eroding fields with deep topsoil layers; rich, productive subsoils; or a combination of both. In some regions, western Iowa and the Palouse in the Pacific Northwest, for example, many soils possess these favorable characteristics. In areas such as the southern Piedmont region that stretches through several mid-Atlantic states, almost all soils are vulnerable to long-term productivity losses because of relatively shallow topsoil layers and subsoils that are unfavorable to root development. The relatively few studies that have focused on a variety of soils in the United States have been helpful in establishing the range of vulnerability to different erosion rates. However, many of these past studies are less useful in assessing contemporary erosion-productivity relationships (see Separating the Effects of Erosion and Technology).

Separating the Effects of Erosion and Technology

Yield observations over time are influenced by several interactive forces. Where soil erosion occurs, the relationship between erosion's adverse effects on productivity and technology's yield-enhancing effects poses a complex research challenge. Yield increases brought about by technological change may not only mask the effects of erosion, but failure to account for the ways that erosion may have influenced the rate and cost of technological advances can lead to inaccurate assessments of damage caused by erosion.

Erosion damage cannot be measured solely as a simple function of yield. A correct assessment requires the clear separation of the projected effects of erosion and technology. For example, crop yield might be increased by technological advances that are independent of inherent soil productivity, such as hybrid seed and improved chemical weed-control systems. To evaluate the effects of erosion in such instances, it is necessary to focus on the impact of erosion damage on attainable yield levels, taking into account the effects of technological advances. The question is: How much higher would yields be with new technology if soil had been conserved?

Even on cropland with deep, friable subsoils where technology boosts yields uniformly throughout the range of yield responses, in spite of the extent of erosion damage to the field, assessment of impaired productivity on land that has benefited from technology requires a measure of erosion damage based on a with conservation versus without conservation comparison of yields (Walker and Young, 1986).

The only valid way to assess erosion effects in regions experiencing substantial technological change—the majority of U.S. agricultural areas—is to compare actual yields and profitability under conditions of contemporary management with those under conditions of controlled erosion. Empirical studies that have attempted to disaggregate erosion from technology effects have demonstrated that yield response to a given new technological advance on conserved soils can be greater than that on eroded soils (Walker and Young, 1986). Research has been initiated in several parts of the country to develop new analytical methods for more accurately quantifying complex erosion-technology interactions.

For many years technology has been thought of as a mask or compensation for erosion. The availability of NRI data has helped analysts recognize that the relationship of erosion and productivity is complicated by technology. Understanding the interactions of erosion, productivity, and technology is a necessary step in reliably estimating the true costs and benefits of erosion-control investments.

Long-Term Productivity

The degree of impact of erosion on long-term productivity varies widely. Many different concepts and experimental designs have been advanced in an effort to quantify these impacts. Proper interpretation of the stated hypotheses and the experimental results is important.

Long-term productivity losses might come about through reduced optimal yields or through higher costs to attain optimal yields. Erosion can also adversely affect a soil's responsiveness to future yield-enhancing technologies. An important question is whether yield increase over time will be greater in the future if less topsoil is displaced by erosion (Walker and Young, 1982). Alternatives must be placed in an economic and social context. A basic goal of soil conservation is to maintain the full potential of the soil to achieve a high level of production at per unit cost over a long period.

Practical and reliable techniques for estimating the long-term economic costs of soil erosion at specific locations are not available. The standard approach of farm-level conservation planning is to identify the physical magnitude of an erosion problem, not its economic consequences, and present the farmer with a series of alternative conservation systems for reducing the volume of soil displacement. The systems vary in cost and effectiveness. Some might be impractical, depending on the farmer's complement of machinery or operation. Ideally, the farmer selects a system that will make his operation more profitable and achieve the greatest reduction in erosion per dollar spent. Choices among conservation alternatives, however, are made without the aid of sound estimates of short- and long-term benefits. Costs, on the other hand, are generally known and are usually significant.

Current research that identifies the effects of erosion on productivity will improve the general scientific understanding of the magnitude and extent of erosion damages. The development of practical methods for field estimation of both short- and long-term erosion damages to productivity would augment current research. Such new methods would be useful to farmers and conservationists.

In developing models and procedures for assessing the effects of erosion on land productivity, the USDA should reconsider the current balance between farm-level basic research on the modeling of erosion processes and more applied, aggregate-level research. (Aggregate-level research is designed to provide technical information to support USDA and state soil conservation activities as well as the conservation planning undertaken by individual farmers.)

The secretary of agriculture should direct a team drawn from the SCS, the Agricultural Research Service, the Extension Service, the Cooperative State Research Service, the Economic Research Service, and other relevant research and administrative organizations to review and periodically report on current research and methods in the area of erosion-productivity interactions. The purpose should be to identify promising techniques for applying such research in the field, to assist extension workers and conservationists in applying research results to farm-level economic planning, and to coordinate future research to enhance its applications in the field and in program development.

Erosion-Productivity Models

Most of the research on the effects of soil erosion on agricultural productivity has been in the form of empirical studies at specific geographic locations (Crosson, 1984; Larson et al., 1983). Results have helped researchers quantify the impact of on-farm erosion damages and have drawn attention to the need for improved soil conservation practices. Empirical studies have also helped researchers identify the soil characteristics and management practices that most significantly influence erosion-productivity damage under many field conditions. However, research methods differ greatly from one study to the next, and the results of these studies are often pertinent to relatively few soils experiencing different rates of soil erosion. As a result, researchers have not been able to generalize about the effects of erosion on productivity for broad geographic areas or to consistently compare effects from one area to another.

Since the release of the 1977 NRI data, however, several mathematical models have been developed that attempt to quantify erosion-induced productivity effects for much broader geographic areas, including the entire United States. Two are discussed here: the Productivity Index (PI) model and the Erosion-Productivity Impact Calculator (EPIC) model. The PI model uses smaller amounts of input data and less technical data. In that sense it is limited, but it is relatively simple and inexpensive to use. The EPIC model is more complex and goes beyond the PI model in attempting to analyze the physical and chemical relationships affecting erosion-productivity. These models and others draw upon the findings of empirical studies of erosion-productivity relationships, data contained in the NRIs, and other data sources. Thus far, the models mainly have been used for policy analysis and program evaluation at the national, regional, and state levels. With further refinement the results of erosion-productivity models should have important application in conservation planning on the farm.

The PI Model

In the late 1970s, researchers at the University of Missouri developed a model of productivity, the PI model (Kiniry et al., 1983). As refined by a research team at the University of Minnesota, the PI model allows comparisons of soil productivity between an ideal soil, in which bulk density, pH, aeration, and available water storage capacity are optimal for root growth, and any other soil for which values for these parameters are available (Pierce et al., 1983). The computerized data base composed of county-level soil surveys, Soils-5, contains such information for thousands of soil profiles. The NRI design permits cross-tabulation with the Soils-5 data base. Information recorded on most NRI sample points can be cross-referenced via computer with detailed Soils-5 information about similar soils.

Once a PI has been constructed for a particular soil profile, changes in productivity that would be expected as a result of erosion can be simulated by evaluating the PI for progressively lower portions of the soil profile—the portions that become the zone of plant root growth as erosion strips away the surface soil. The depth of the soil, its quality as a medium for plant growth, and the amount of erosion (rate multiplied by time) become the determinants of future soil productivity.

The PI model assumes that the availability of plant nutrients will not be a limiting factor on crop production. Damage estimates are generally conservative, because the model assumes that any direct reduction in nutrient levels or nutrient storage capacity sustained from erosion can be overcome by the addition of increasing amounts of fertilizer.

Table 4-1 shows some characteristics and PI model results for land in

TABLE 4-1 Relationships Between Erosion and Measures of Soil Productivity in Six MLRAs for Land Planted in Row Crops

	Rates of Erosion (tons/acre·year)			Measures of Productivity Impact		Change in PI over 100 years (percent)
MLRA	Potential	Actual	Tolerable[a]	PI[b]	V[c]	
105	71.6	11.4	4.7	0.84	0.23	5.6
109	45.1	15.2	3.9	0.79	0.17	6.9
113	25.4	8.0	3.4	0.72	0.21	4.4
103	11.2	4.1	4.9	0.88	0.27	1.9
108	28.9	8.8	4.9	0.91	0.16	2.4
115	32.0	10.2	4.6	0.83	0.14	3.6

[a] Soil loss tolerance (T) value.
[b] Productivity index: Ratio of soil productivity after an increment of erosion in contrast to uneroded state. (PI does not equal 100 for any soils in the initial period.) Approximates loss in productivity from erosion.
[c] Vulnerability value: Approximates the rate of change over time for a given soil in the PI value. High V values indicate relatively high susceptibility to erosion-induced productivity loss.
SOURCE: 1982 NRI and Soils-5; adapted from Runge et al., 1986.

six Major Land Resource Areas (MLRAs) that were reported as planted to corn or soybeans in the 1982 NRI. In general, MLRAs 103 (the central Iowa and Minnesota till prairies), 108 (the Illinois and Iowa deep loess and drift), and 115 (the central Mississippi Valley wooded slopes), which have fairly low average potentials for sheet and rill erosion (RKLS product values), exhibit high average levels of soil productivity (PI values).

The V value approximates the rate of reduction in the PI value for uniform, incremental reductions in soil depth resulting from erosion. It is an indication of the average vulnerability of the land to erosion in these MLRAs (Pierce et al., 1984). Whereas the PI value is a static measure of productivity, the V value is an indicator of how rapidly a soil's productivity can be reduced by erosion. Vulnerability increases with increasing V values. Table 4-1 indicates that there are exceptions to the general observation that high rates of erosion are associated with significant soil productivity damages. MLRA 103 has the lowest potential for erosion of the six MLRAs listed in Table 4-1 and a fairly high productivity rating, but it is also the MLRA most vulnerable to a given amount of erosion, as indicated by the higher V-factor value for the soil.

Uses of the PI Model Some researchers have suggested that V values, or similar measures of a soil's vulnerability to erosion damage, might be applied to refine conventional soil loss tolerance limits (Pierce et al., 1984; Runge et al., 1986). The committee believes that such approaches should be used to improve the conceptual and empirical basis of T values with respect to erosion-productivity relationships. This could be accomplished within a reasonable time and at acceptable developmental and implementation costs.

Analytical tools such as the PI model can be used to investigate the ways that soil erosion influences productivity across the country over time. Table 4-1 shows the predicted effects on the average PI over 100 years for six MLRAs, assuming that the average sheet and rill erosion rates reported in the 1982 NRI continued. Again, the predicted decline in productivity incorporates the assumption that damage to the plant nutrient regime, or any weed or pest problems exacerbated by erosion, will be corrected through adjustments in input levels, management decisions, or both.

MLRAs with high inherent potential for erosion as well as high actual erosion rates—MLRAs 105 (the northern Mississippi Valley loess hills), 109 (the Iowa and Missouri heavy till plain), 113 (the central claypan areas of Missouri and Illinois), and 115—would sustain the most serious productivity damages. MLRAs 105, 109, and 113 exhibit greater susceptibility to damage, in some cases more than twice the average decline in PI. However, the average reduction in PI value is weighted heavily by modest PI reductions in MLRAs 103 and 108, which together account for more than 60 percent of the acreage evaluated in the table.

Geographic Variation Broad geographic assessments of erosion-productivity damages, whether for the entire United States or smaller aggregations, provide useful indications of overall conditions. But these aggregations can obscure important variations observed among regions or in local areas within regions such as MLRAs. Erosion-productivity problems are not uniformly distributed across the country, within regions, or in individual fields of cropland. Assessments of erosion damages that emphasize average reductions in crop yields or other measures of productivity can underplay the more serious, geographically concentrated instances of severe erosion-induced productivity damages. The committee believes that the highly variable nature of erosion-related productivity should be more systematically considered in studying erosion and in the design and implementation of erosion control programs.

The EPIC Model

The Resource Conservation Act (RCA) mandates that the USDA perform an appraisal of soil and water resource conditions and trends every five years. As part of the 1985 RCA appraisal, an interagency research group involving the SCS, the Economic Research Service, and the Agricultural Research Service developed a mathematical model for estimating erosion impacts on crop yields and costs of production (in the form of additional requirements for such production inputs as fertilizer and water). The EPIC model is a sophisticated descendent of the Yield Soil Loss Simulator, the model used to evaluate erosion-productivity effects in the 1980 RCA appraisal. The model draws upon NRI erosion estimates to develop baseline conditions and can simulate erosion rates and long-term productivity effects that would be expected from alternative cropping and conservation management conditions.

In view of the importance of testing the interactive effects of technology and erosion damage on crop yields and other measures of productivity, the EPIC model, used in conjunction with erosion information from the NRIs, is a potentially useful tool for scientific research and policy analysis. The EPIC model is more complex and demanding of data than the PI, and like the PI model, EPIC will provide insights into the erosion-productivity relationship. The committee notes that until the full impacts of erosion on all factors contributing to agricultural productivity can be quantified, it is likely that the short- and long-term agronomic and economic impacts of erosion will continue to be poorly understood.

Soil Erosion and Water Quality

The most serious damages of erosion occur, in some cases, after sediment and runoff have left the field. A review of offsite erosion damages conducted by the Conservation Foundation estimates that the social cost of these damages may have totaled $3 billion to $13 billion in 1980 alone, excluding biological damages (Clark et al., 1985). Damages appraised in the study include impaired recreation and damage to water storage facilities, navigational impacts, property values, and commercial fisheries. Also included are damage to water conveyance systems, water treatment facilities, municipal and industrial facilities, and flooding.

Field studies and monitoring activities cited earlier (see Chapter 2) suggest the serious nature of offsite erosion problems. More recently, attention has been focused on the potential impacts of agricultural

Spring runoff on a plowed field, Brown County, Wisconsin. Such runoff often transports sediments, animal wastes, plant nutrients, and pesticides—all causes of nonpoint water pollution. Credit: U.S. Department of Agriculture, Soil Conservation Service.

practices on groundwater quality, particularly the use of fertilizers and pesticides. The committee believes that strategies for protecting the quality of surface water and groundwater will have to consider the possibility that groundwater pollution can be aggravated by techniques intended to mitigate erosion problems. Conservation practices often rely more heavily on agricultural chemicals than conventional management and at the same time increase infiltration of water, which may aggravate pollution of groundwater.

Potential Uses of NRI Data

There are major gaps in the scientific data on the basic processes of offsite damages—and pollution of surface water and groundwater—caused by erosion and agricultural management practices. These problems may be serious in many parts of the country.

Currently, the usefulness of NRI data for assessing offsite erosion damages is hampered by the lack of basic scientific understanding of the effects of soil erosion and runoff on surface water and groundwater quality. The Universal Soil Loss Equation (USLE), for example, was not designed to evaluate sediment delivery into water bodies or other nonpoint pollution problems. NRI estimates of sheet and rill erosion do provide an indication of potential sediment loads from agricultural land. The difficulty is in translating these potential loads into actual sediment levels observed in watercourses and then assessing the damage caused by the added sediment.

NRI values for select physical factors of the USLE can be helpful in analyzing surface water quality when properly integrated into mathematical models such as those mentioned in Chapter 3. These models focus on the linkages between erosion, soil, and chemicals carried by runoff at the field's edge and additional changes that occur as runoff of water and sediment moves into or through small streams. Additional work is needed to better define the hazards posed by infiltration of nutrients or chemicals into groundwater in diverse regions that are subject to a variety of agricultural practices (Crosson and Brubaker, 1982) and hazards associated with sediments and chemical constituents that are transported and deposited in water systems.

As noted earlier, the committee believes that sufficient evidence exists to conclude that environmental hazards and social costs associated with water pollution derived from agricultural lands are significant. Improved control measures and strategies are needed; better information is essential to devise such strategies. As such information is developed, current and new data from the NRI can contribute to the assessment of the offsite impacts of agricultural activities.

5

Assessing Conservation Practices and Land Classification Schemes

One of the most important uses of the NRIs is to analyze the diversity of erosion conditions across the country. The inventories are also used to determine the effectiveness of conservation practices in mitigating erosion problems. This chapter describes new information on the distribution of conservation practices that were recorded in the 1982 NRI. It also focuses on the analytical component shared by most new policy initiatives—the classification of cropland according to its susceptibility to erosion-induced damages. Such damages may affect the land itself or water bodies to which it is tributary. Accurate identification of croplands especially susceptible to erosion is dependent upon a sound land classification scheme.

Conservation Practices

The 1982 NRI contains data on the conservation practices employed by farmers on all types of land. The survey was designed to record up to three practices per acre. Many acres are treated with combinations of practices: terrace systems in conjunction with contour farming, and conservation tillage with stripcropping, for example.

The 1982 NRI results support three basic conclusions regarding conservation practices that first emerged from analysis of the 1977 NRI (U.S. Department of Agriculture, 1981). First, nearly 50 percent of the intensively cultivated cropland in the United States is treated with some conservation practice. Second, many of the practices currently in

Miles of grass-backed, tile outlet terraces conserve soil and moisture on this farm in Montgomery County, Iowa. Conservation tillage is also used on the farm to control erosion. This is an example of a highly effective conservation system involving C- (cover and management) and P- (supporting soil conservation practices) factor practices in accordance with a conservation plan. Credit: U.S. Department of Agriculture, Soil Conservation Service.

use—again, about half—are used on land not subject to excessive erosion losses. Third, much of the land most in need of erosion control—as defined by the USLE—is not treated with any practice.

Table 5-1 shows the distribution and frequency of application of conservation practices according to ranges of erosion potential. Ideally, the frequency of application should increase as inherent erosion potential increases. This does not appear to be the case. For the nation as a whole, the percentage of acres treated with one or more conservation practices appears to decline with successively higher potential erosion. A slight increase is suggested for the Corn Belt, Iowa, and the highest potential erosion in Georgia. The percentage of acres treated appears relatively constant across the range of potential erosion. Practices on land with erosion potential less than 10 have probably been adopted for reasons other than erosion control, such as fuel and labor savings.

The acreage treated with conservation tillage methods is increasing (see Chapter 1). The purpose of such tillage practices is to leave crop residues sufficient to cover a minimum of 30 percent or more of the soil

TABLE 5-1 Acreage Treated with Conservation Practices in the United States, the Corn Belt, and Selected States[a]

Erosion Potential (tons/acre·year)	Total Acreage (millions)	Acres (millions) Treated with Given Number of Practices					One or More (percent)
		None	One	Two	Three	One or More	
United States							
0–<10	220	92	82	32	15	129	58
10–<20	93	44	30	12	7	49	53
20–<40	55	27	16	7	5	27	50
40–<60	20	11	6	2	1	9	47
60–<100	17	9	5	2	1	8	47
100–<150	8	4	3	0.9	0.4	4	47
>150	7	4	2	0.8	0.3	4	47
Corn Belt							
0–<10	48	26	17	4	1	22	46
10–<20	25	14	7	2	1	11	44
20–<40	18	9	5	2	2	8	48
40–<60	8	4	2	0.9	0.8	4	49
60–<100	8	4	2	1	0.7	4	50
100–<150	4	2	1	0.6	0.3	2	49
>150	4	2	1	0.5	0.2	2	51
Georgia							
0–<10	2	1	0.4	0.1	0.08	0.6	30
10–<20	3	2	0.4	0.2	0.09	0.7	28
20–<40	1	1	0.2	0.1	0.03	0.4	28
40–<60	0.4	0.3	0.1	0.04	0.01	0.1	36
60–<100	0.3	0.2	0.07	0.01	0.01	0.1	34
100–<150	0.08	0.06	0.02	0.002	0.001	0.03	30
>150	0.04	0.02	0.02		0	0.02	39
Iowa							
0–<10	9	4	4	1	0.2	5	56
10–<20	6	3	2	0.7	0.2	3	50
20–<40	4	2	1	0.6	0.2	2	50
40–<60	2	1	0.7	0.4	0.1	1	50
60–<100	3	1	0.8	0.5	0.2	2	66
100–<150	2	0.6	0.5	0.3	0.1	1	50
>150	2	0.6	0.4	0.3	0.1	1	50
Ohio							
0–<10	6	3	3	0.4	0.08	3	52
10–<20	3	2	0.8	0.2	0.02	0.1	35
30–<40	2	1	0.4	0.08	0.02	0.5	33
40–<60	0.6	0.4	0.2	0.05	0.01	0.2	40
60–<100	0.5	0.3	0.1	0.05	0.005	0.2	39
100–<150	0.3	0.2	0.1	0.03	0.01	0.1	40
>150	0.4	0.2	0.1	0.04	0.005	0.2	46

[a]States were selected to illustrate regional differences. In addition, the numbers in this table have been rounded for the convenience of the reader. The precise numbers generated from NRI data are the statistical summations of all acreage represented by sampling points; they should be used for further technical analyses.

SOURCE: 1982 NRI.

surface. Erosion control benefits are proportional to the degree of cover left on the surface.

Figure 5-1 shows the variability in the degree of protection afforded by conservation tillage practices. Sixty-six percent of the 100 million acres treated with conservation tillage in 1982 had C-factor values between 0.1 and 0.30. However, a considerable amount of the acreage reported in conservation tillage had C-factor values that were as high or higher than those expected under conditions resembling continuous plantings of corn, with the land plowed in the fall or spring. This finding reflects the diversity of practices that can be classified as conservation tillage, some of which leave limited residue cover. Erosion control benefits on such lands are minimal. It also suggests a possible need to more carefully define and use the term conservation tillage.

The committee believes that improved field estimates of surface residue cover should be used in future NRIs. Surface cover, incorporated crop residue (that near the surface), and roughness factors can be estimated and have a major effect on C-factor values.

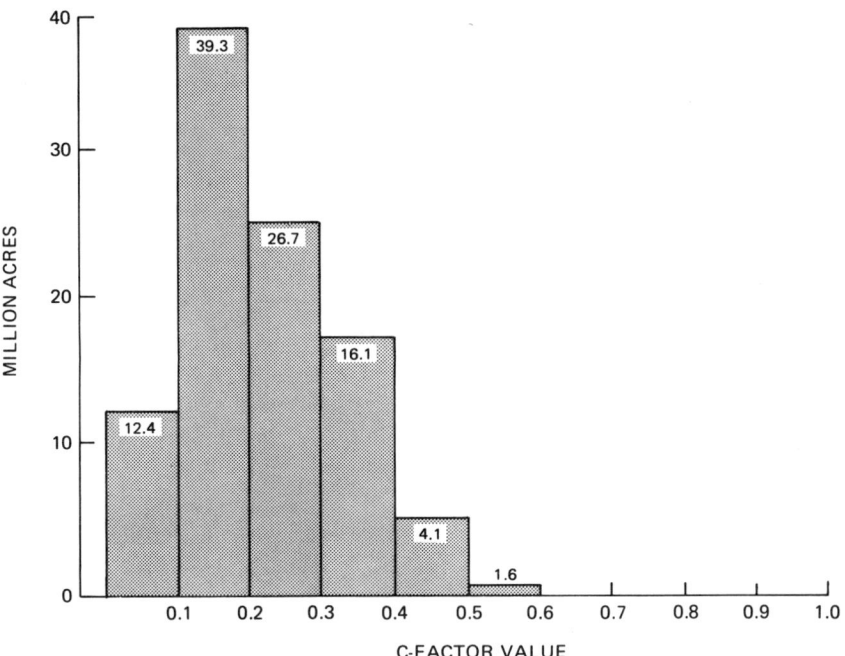

FIGURE 5-1 Distribution of C-factor values on cropland treated with conservation tillage, 1982.

Regional variations in the distribution of conservation tillage practices are shown in Table 5-2. For example, MLRA 103, located in southern Minnesota and northern Iowa, contains about 14 million acres of cropland. Of that, about 13 percent was in conservation tillage in 1982. About 70 percent of the cropland had an RKLS value of 10 tons/acre·year or less, a very small potential for sheet and rill erosion. About 12 percent of these nonerodible or slightly erodible acres were treated with conservation tillage.

Ideally, the percentage of acres in a given MLRA treated with conser-

TABLE 5-2 Four MLRAs: Distribution of Acreage by Crop, Potential Erosion Class, and Conservation Tillage

Crop and MLRA	Cropland (1,000 acres)	Conservation Tillage (% acres)	Percent Acres in Potential Erosion Class and in Conservation Tillage[a] (RKLS in tons/acre·year)		
			< 10	10–20	> 20
MLRA 103					
Corn	7,162	14	69 (13)	18 (16)	13 (18)
Soybeans	5,854	13	72 (12)	16 (16)	12 (16)
Wheat	388	10	76 (11)	17 (7)	8 (3)
Cropland	14,443	13	70 (12)	17 (15)	13 (16)
MLRA 105					
Corn	3,166	28	13 (19)	15 (25)	72 (30)
Soybeans	198	15	31 (5)	21 (19)	47 (20)
Wheat	*	*	*	*	*
Cropland	5,820	23	11 (14)	12 (23)	77 (24)
MLRA 134					
Corn	445	21	6 (14)	29 (15)	65 (24)
Soybeans	4,293	16	4 (9)	44 (13)	52 (19)
Wheat	552	23	2 (12)	33 (21)	65 (25)
Cotton	860	4	7 (3)	43 (7)	50 (3)
Cropland	7,469	14	4 (7)	44 (13)	52 (17)
MLRA 136					
Corn	804	12	7 (8)	20 (20)	73 (10)
Soybeans	1,149	18	3 (12)	15 (11)	82 (20)
Wheat	744	18	3 (56)	14 (11)	83 (18)
Cropland	4,223	12	4 (17)	15 (11)	81 (12)

[a] Percent conservation tillage is given in parentheses.
*Number of sampling points too small.
SOURCE: Pierce et al., 1986.

The soil is relatively undisturbed when the no-till method of conservation tillage is used (Montgomery County, Maryland). The field will be sprayed for grass and broadleaf weeds after planting. Erosion will be reduced to less than one-tenth the rate expected from a comparable field under conventional tillage. Credit: U.S. Department of Agriculture, Soil Conservation Service.

vation tillage should increase significantly in land groups with progressively higher inherent erosion potential. This is not the case in MLRA 103; it is true in MLRA 134 (the southern Mississippi Valley silty uplands of Mississippi, Tennessee, and Kentucky), where the proportion of land treated with conservation tillage increases substantially from 7 percent in the lowest RKLS groups to 17 percent in groups with RKLS values greater than 20. These and other regional variations demonstrate the existence of widely differing rates of adoption of conservation tillage technologies on erosion-prone cropland.

Several studies based on 1977 NRI data indicated that much of the land in conservation tillage had a modest potential for sheet and rill erosion before the practice was adopted (American Farmland Trust, 1984). Two-thirds of the land treated with conservation tillage in the Corn Belt in 1977 had an inherent sheet and rill erosion potential of less than 20 tons/acre·year, which is below the national average RKLS value of 21.7 tons/acre·year reported for all cropland.

Concentration of Sheet and Rill Erosion on Cropland

One of the most significant practical applications of the NRI is likely to be new research leading to further documentation of the extent and geographical characteristics associated with the concentration of excessively erodible land on a relatively small portion of the land base.

The concentration of sheet and rill erosion was pronounced on land used to produce close-grown crops in 1982. About 84 percent of that land (97.6 million acres) had sheet and rill erosion rates less than 5 tons/acre·year; only about 3 percent had rates greater than 15 tons/acre·year. It is important to note, however, that much of the acreage planted to close-grown crops, wheat in particular, is in areas where wind erosion is the chief soil conservation problem. Nevertheless, land in row and close-grown crops that erodes at rates above 15 tons/acre·year accounts for more than 40 percent of the total sheet and rill erosion of land in those uses.

Table 5-3 illustrates the relatively large proportion of cropland with modest potential for sheet and rill erosion. About 277 million acres of cropland and 133 million acres of the land used for row crops—about two-thirds of the total acreage in these uses—had an inherent potential for sheet and rill erosion of less than 15 tons/acre·year. Under average farming conditions (C-factor value of 0.30, P-factor value of 0.91), sheet and rill erosion rates would average less than 4 tons/acre·year on this land. The remaining 140 million acres of cropland had an inherent potential erosion rate of more than 15 tons/acre·year; about 52 percent of that acreage was used for row crops.

TABLE 5-3 Cropland Uses by RKLS Factor, United States, 1982 (million acres)

Potential Erosion, RKLS (tons/acre·year)	Row Crops	Close-Grown Crops	Hay	Other Crops	Total
0–<5	44.8	38.0	12.0	19.4	114.3
5–<10	56.7	28.0	6.5	13.5	104.8
10–<15	31.6	16.6	3.0	6.7	58.0
15–<20	18.1	9.3	2.6	4.1	34.1
20–<20	10.8	6.1	1.5	2.4	20.8
25–<30	7.3	4.2	1.4	1.8	14.7
30–<35	5.4	2.9	1.0	1.3	10.5
35–<40	4.0	1.9	1.0	1.0	7.8
40–<50	6.0	2.6	1.4	1.5	11.5
50–<75	8.9	3.2	2.1	2.3	16.6
75–<100	4.7	1.4	1.4	1.3	8.8
>100	7.9	1.6	3.4	2.7	15.6
Total	206.3	115.6	37.5	58.1	417.5

SOURCE: 1982 NRI; adapted from Rosenberry and English, 1986.

The committee has given special attention to lands with extremely high potential rates of sheet and rill erosion (50 tons/acre·year or more). About 41 million acres, or 9.8 percent of total cropland, falls into this highly erodible category. On lands susceptible to high rates of sheet and rill erosion, it is generally difficult and costly to devise effective conservation farming systems.

Implications for Policy and Program Administration

The 1982 NRI has advanced the understanding of the concentration of soil erosion and the distribution and effectiveness of soil conservation practices. The papers included in Volume 2 of *Soil Conservation: Assessing the National Resources Inventory* are representative of analytic capabilities that are possible using the 1982 NRI data. Their major conclusions have important implications for government policy and programs and for the design of on-farm conservation strategies and the mitigation of adverse offsite effects of erosion.

Because of state and federal budget constraints, new public and private initiatives are needed to enhance the cost-effectiveness of soil conservation investments.

Targeting It is probable that traditional USDA conservation programs will receive significantly less funding in the future. Budgets for technical assistance and cost-sharing programs will probably remain near current levels in nominal dollars.

Some states are considering ways to compensate for reduced federal investment in conservation. For example, Missouri voters recently enacted a special tax to generate income for conservation cost sharing (Johannsen, 1986). Presentations based on analysis of data from the 1977 NRI were instrumental in convincing the Missouri legislature of the merits of this special tax.

More reliable techniques should be developed for targeting public and private soil conservation investments according to the potential to affect on-farm productivity, offsite damages stemming from soil erosion, or both. In particular, future targeting directed to onsite damages increasingly should be based on indices of the relationship between erosion and productivity. Models such as the Productivity Index (PI) and the Erosion-Productivity Impact Calculator (EPIC) have the potential to quantify more precisely the physical and economic effects of erosion damage and identify land that is acutely vulnerable to erosion damage. Application in the field will require an appropriate data base that includes information from soil surveys.

Soil conservation activities, public and private, should be systematically targeted or concentrated—much like erosion—toward those lands that are most susceptible to soil erosion damage or that contribute most significantly to serious offsite pollution problems. The committee recommends that the USDA expand the scope and shorten the timetable of targeting soil conservation programs toward the most fragile cropland, rangeland, forestland, and pastureland. Under current fiscal constraints it is likely that, to be effective, targeting will need to be more selective than previously experienced in the conservation field. If erosion is to be controlled, long-term land diversion programs designed to convert highly erodible cropland into stable forage- or forestry-based land uses are needed.

Erosion Reduction Goals Estimates of soil erosion are imperfect indicators of actual on-farm or off-farm damage. A better understanding of erosion processes is clearly desirable to enhance the ability to quantify and predict the effects of different forms of erosion on soil productivity and environmental quality. At the same time, increases in the sophistication of predictive models—and the data bases necessary to apply them—often come at exponentially higher costs.

The committee believes that the practical conservation benefits likely to result from expanded data sets and new modeling exercises should be more systematically appraised in establishing budgetary priorities. The sophistication and cost of some models have already become prohibitive for national applications. Identification of the appropriate mix of research on the fundamental process of erosion and the role of improved models and data for future conservation policy and programs should be more clearly articulated to maximize benefits from limited funds.

Modeling and data needs will change as the scientific understanding of conservation needs changes. In the last few years, for example, optimism over the success of conservation tillage in controlling erosion in many parts of the United States has been tempered by concern and supported by very limited information suggesting that conservation tillage, coupled with common fertilizer and pest management practices, might increase the level of pollutants entering ground and surface waterways. In response, however, considerable research is currently under way to reduce the reliance on chemicals in conservation tillage.

Careful analysis is needed to ensure that future NRIs, new empirical models, and new methods will provide the information necessary to meet the dual challenge of soil and water conservation, both qualitatively and quantitatively. The committee believes that reducing erosion and enhancing soil productivity while conserving water and protecting water quality in the broadest sense must become the dominant objective in soil conservation policy.

The USDA and state conservation agencies should continue to emphasize gross erosion reduction goals and accomplishments in program evaluation, design, and administration. Whenever possible, erosion reduction goals relating to productivity should be redefined in accordance with available, reliable indicators of a soil's susceptibility to erosion damages.

Recent work on productivity indices and the ways that erosion may influence potential productivity losses over time is beginning to provide more sound quantitative measures of the impact of erosion. The complexity and variability of the processes were noted in Chapter 3. The ratio of inherent erosion potential to estimates of soil loss tolerance limits proposed by the USDA Fragile Soils Work Group has the merit of recognizing the importance of including both erosion and a measure of productivity in an indicator of damage potential. At the same time, uncertainties inherent in the estimates of T values remain, and the ratios are surrogates rather than measures that attempt to capture the impact of erosion on potential productivity over time. With continuing research, these processes can be further refined. They should be made an integral part of USDA program evaluation and management as soon as possible.

The Land Capability Class System

For many years soil information has been collected for most of the intensively cropped regions in the country through the National Cooperative Soil Survey Program. Most of the acreage surveyed was assigned to one of eight classes and four subclasses in the SCS Land Capability Class System (LCCS), following subjective criteria and the judgment of regional experts. The LCCS classes of land range from best (class I) to most limited for agricultural production (class VIII). The numbers designate the severity of the problem for crop uses. The letters e, w, s, and c indicate whether the problem is caused by erodibility; wetness; stoniness, shallowness, or drought; or climate, respectively.

Like any classification system, the LCCS was designed to satisfy specific objectives. The variables selected and the range of classes were dictated by the objectives and the scale of intended use. Interest was focused on a number of factors related to agricultural production. The LCCS is a valuable tool, but it was not designed to provide quantitative estimates of either actual or potential erosion rates. In addition, when this system is used, it is often difficult to distinguish the physical characteristics of the land at the time of mapping from the effects of management practices that have been used in previous years. The committee does not focus on the broad, possible uses of the LCCS, but only on limitations of the LCCS specifically related to the delineation of erosion and alternative approaches that might better meet this objective.

By using the erosion data in the 1977 and 1982 NRIs, particularly the USLE information, land management practices can be taken into account. Analyses of 1977 and 1982 NRI data raise questions about the utility of the LCCS for applications involving the calculation of erosion rates before or after applications of soil conservation practices. The LCCS has been used for more than 30 years; alternative schemes might be better suited to contemporary applications concerned with land erosion and conservation management.

Factors of Inconsistency

Several factors probably contribute to LCCS incongruities with respect to inherent potential and actual erosion rates. The system was essentially devised and implemented prior to the development of quantitative methods of estimating erosion in the field. Another important source of variation within the system is the local, subjective nature of LCCS determinations by SCS field personnel. When the system was developed 50 years ago, there was no need for categories of land groupings that would be consistent across the country. However, the marked variation in erosion conditions, even within localized areas such as MLRAs, remains today.

The LCCS was devised to categorize land according to its physical characteristics, without taking into account management practices such as conservation tillage in use at the time of classification. For the purposes at hand, the LCCS could be improved by reclassifying soils according to estimates from the soil erosion equations. This could be accomplished by basing reclassification on the USLE and Wind Erosion Equation (WEE) factors that reflect unchanging climate and soil characteristics: the RKLS product for sheet and rill erosion and the I, C, and L factors for wind erosion. The NRIs provide very useful information for updating LCCS classifications and developing alternative classification schemes.

Rates of Erosion: A Basis for Classification

The observation that suggests a need for a new land classification scheme emerged from analysis of the distribution of erosion rates on land classified within given LCCS land classes and subclasses, which was reported by Heimlich and Bills (1986). They found that there is a wide variation of erosion and inherent erosion potential on lands classified within the same class and subclass. This is true for most regions and throughout the country. Moreover, land categorized within specified ranges of potential erosion—with RKLS products between 20 and 30, for example—also is often erratically distributed across several different land classes and subclasses in the LCCS.

For example, about 59 million acres of the total cropland acreage were classified in LCCS class IIIe in the 1982 NRI (Table 5-4). Cropland classified as IIIe is generally considered suitable for intensive cultivation, with appropriate conservation measures. Surprising variation, however, is found in NRI values for the inherent potential for sheet and rill erosion on class IIIe land nationwide. Land within that class (and others) differs substantially in its vulnerability to erosion.

At one extreme, class IIIe cropland includes about 21 million acres that have an inherent erosion potential (RKLS product) of less than 10 tons/acre·year, according to 1982 NRI data. Wind or ephemeral gully erosion might be significant problems on some of this land; however, sheet and rill erosion are not. Under average management conditions (represented by a C-factor value of 0.30) average annual erosion rates of less than 2 tons/acre·year would be expected.

This class IIIe acreage with a low erosion potential contrasts dramati-

TABLE 5-4 Distribution of Acreage with Inherent Erosion Potential by Capability Class, According to Erosion Potential[a]

Erosion Potential, RKLS (tons/acre·year)	Capability Class, Row and Close-Grown crops (million acres)												
	I[b]	IIe	IIIe	IVe	VIe	VIIe	IIw	IIIw	IVw	IIs	IIIs	IVs	VIs	VIIs	Others

Erosion Potential	I[b]	IIe	IIIe	IVe	VIe	VIIe	IIw	IIIw	IVw	IIs	IIIs	IVs	VIs	VIIs	Others
United States															
0–<10	18	24	21	7	1	0.07	40	16	3	10	7	3	0.7	0.2	16
10–<20	7	24	9	3	0.7	0.02	14	10	1	3	1	0.6	0.3	0.09	2
20–<40	1	19	2	3	1	0.04	3	1	0.2	0.8	0.4	0.3	0.2	0.07	0.2
40–<60	0.06	5	6	2	0.5	0.04	0.3	0.2	0.06	0.05	0.06	0.1	0.07	0.04	0.03
60–<100	0.02	2	7	3	0.7	0.05	0.1	0.07	0.07	0.006	0.03	0.07	0.04	0.03	0.006
100–<150	0.003	0.3	3	2	0.6	0.07	0.03	0.01	0.005	0.002	0.004	0.006	0.02	0.02	0.002
>150	0.001	0.09	1	2	1	0.3	0.01	0.007	0.005	0.001	0.001	0.009	0.03	0.04	0.002
Total	26	73	59	21	6	0.6	58	28	5	14	8	4	1	0.5	18
Corn Belt															
0–<10	9	3	0.5	0.07	0.02	0.001	24	50	0.3	2	0.6	0.5	0.04	0.006	0.3
10–<20	3	10	1	0.2	0.04	0.004	7	1	0.03	0.8	0.2	0.2	0.03	0.006	0.07
20–<40	0.4	9	4	0.6	0.08	0.006	1	0.2	0.02	0.3	0.04	0.1	0.04	0.005	0.01
40–<60	0.03	2	4	0.7	0.1	0.003	0.2	0.03	0.04	0.02	0.005	0.03	0.02	0.01	0.002
60–<100	0.01	0.7	4	1	0.2	0.02	0.06	0.02	0.06	0.002	0.002	0.008	0.02	0.02	0
100–<150	0.003	0.1	2	0.1	0.3	0.02	0.02	0.005	0.004	0.001	0.002	0	0.01	0.01	0.001
>150	0.001	0.03	0.8	1	0.5	0.1	0.01	0.003	0	0.001	0	0	0.009	0.02	0.001
Total	12	25	16	5	1	0.2	32	6	0.5	3	0.9	0.7	0.2	0.07	0.3

[a]The numbers in this table have been rounded for the convenience of the reader. The precise numbers generated from NRI data are the statistical summations of all acreage represented by sampling points; they should be used for further technical analyses.
[b]Sum of all possible subclasses in Class I.

SOURCE: 1982 NRI.

cally with another 11 million acres, also classified IIIe, that has an RKLS value of more than 60 tons/acre·year. Farmed under average conditions, this land—18 percent of the total IIIe class—would erode at rates of 15 tons/acre or more annually.

U.S. cropland classified as IIIe would exhibit large variations in sheet and rill erosion rates under comparable cropping and management conditions. The climatic and physical characteristics that affect sheet and rill erosion, as reflected by the RKLS product, are not homogeneous within the LCCS class IIIe nationwide. Similar variability in inherent sheet and rill erosion potentials is observed within all land capability classes and within most subclasses.

The committee considered the hypothesis that wind and ephemeral gully erosion problems might explain why cropland with low RKLS values might be classified as IIIe. The NRIs offer no insights into the ephemeral gully erosion question, because the inventories do not contain estimates of that form of erosion (see Chapter 3). The committee believes that it is unlikely that ephemeral gully erosion accounts for much of the classification discrepancy, because the severity of ephemeral gully erosion also appears to be correlated with slope gradient, slope length, and rainfall—critical factors in the USLE.

However, the presence of wind erosion problems on class IIIe cropland can be investigated through the NRIs. Analyses of the 1977 NRI indicate that wind erosion often is a problem on land that has a low RKLS value and is located mainly in the western United States (American Farmland Trust, 1984). The critical USLE factor in this combination is R (rainfall), the value of which is very low in areas prone to heavy wind erosion.

Regional Analyses

Wind erosion probably plays a minimal role in accounting for LCCS inconsistencies according to actual rates of erosion; the inconsistencies are also evident in several MLRAs where wind erosion has never been a significant problem.

Comparisons of the highly erodible acres in a given region by subclass show similar inconsistencies. For example, serious erosion problems exist in MLRA 136, the southern Piedmont of the southeastern United States, when row and close-grown crops are produced on land that has sheet and rill erosion rates in excess of 60 tons/acre·year. Yet, as indicated in Table 5-5, more land with an RKLS product greater than 60 tons/acre·year is classified as class IIe than as class IVe (185,000 compared with 109,000 acres, respectively). Subclasses IIe and IIIe com-

TABLE 5-5 Distribution of Inherent Erosion Potential in the Southeastern Piedmont, MLRA 136

Potential Erosion, RKLS (tons/acre·year)	Capability Class, Row and Close-Grown Crops (100 acres)											
	I[a]	IIe	IIIe	IVe	VIe	VIIe	IIw	IIIw	IVw	IIs	IIIs	IVs	Others
0–<10	119	279	30	0	0	0	489	424	47	0	68	0	10
10–<20	122	3,411	556	69	5	0	486	727	103	107	66	7	15
20–<30	0	5,196	1,332	94	0	0	199	73	19	0	49	0	16
30–<40	0	3,565	1,061	284	56	0	14	0	0	11	0	56	0
40–<60	0	3,823	2,131	772	72	12	13	22	0	11	25	49	0
60–<100	0	1,857	2,143	1,094	327	41	0	26	0	0	0	97	13
>100	0	258	783	897	536	86	0	0	24	0	0	22	67
Total	241	18,389	8,036	3,210	996	139	1,201	1,272	193	129	206	231	121

[a] Sum of all possible subclasses in class I.

SOURCE: 1982 NRI.

bined have almost four times as much row and close-grown cropland in the 60 tons/acre·year and over RKLS group than is reported in class IVe.

Recent SCS staff reports have investigated the usefulness of the LCCS to define erosion potential on cropland with data from the 1982 NRI for 17 MLRAs. One study (Lee and Goebel, 1984) divided 97.5 million acres of cropland into four groupings, following the RKLS-based system developed by Heimlich and Bills (1986). The analysis found considerable variation in RKLS values for individual land capability classes. For example, about 64 percent of the class IIIe cropland acreage in the 17 MLRAs studied (about 10 million of 15.8 million acres) had an inherent erosion potential exceeding 50 tons/acre·year and estimated erosion rates greater than 5 tons/acre·year. About 4.7 million acres (80 percent) of the 5.9 million acres classified as IVe fit this category. About 18 percent of the class IIe cropland (3.9 million acres) also fit this RKLS-based definition of highly erodible. Thus, the definition highly erodible based on the LCCS would inappropriately classify a considerable portion of cropland. Some land that belongs in this category would be excluded, and some land with effective erosion control would be included.

These observed incongruities do not seriously compromise the usefulness of the LCCS as a tool for farm-level conservation planning. A major value of the system is that the subclass e designation signals the landowner and the conservationalist that there is a need for some type of soil conservation treatment. Typically technicians from the SCS or local conservation districts present the land owner with conservation options, tailored to specific fields, that would control erosion to varying degrees at varying costs. This type of planning usually involves estimation of the sheet and rill erosion rates that would be expected before and after the use of select conservation practices. Accordingly, conservation planning generally reflects particular field conditions, regardless of the existing LCCS classifications.

Changes in Land Use

Based on the committee's review of data on the inherent erosion potential of cropland and the distribution of conservation practices, it is clear that new policies and programs deserve study. In addition, new initiatives might be needed to discourage conversion of more erosion-prone cropland to cultivated uses. The committee believes that more attention should be directed toward encouraging desirable land use changes in shaping cost-effective future policies. The maintenance of permanent vegetative cover on land with a high potential for sheet, rill, and wind erosion should figure prominently in future program design, especially during periods of surplus production.

Because of the concentration of soil erosion on the acreage supporting major crops, including wheat, feed grains, soybeans, and cotton, the committee suggests that permanent vegetative cover be used as a tool for conservation strategy. As shown earlier in Table 5-3, a relatively small portion of cropland has very high levels of inherent potential for erosion. Effective conservation systems on such lands are often prohibitively expensive; thus, the land may be unsuitable for intensive crop production—from a resource management perspective.

Furthermore, such land generally provides lower and more variable yields, even if well-designed soil conservation and other management practices are applied. Recent analysis of pastureland, hayland, and other land converted to cultivated crop uses between 1979 and 1981 indicates that about 19 percent of this land (about 2.1 million acres) is highly erodible (Heimlich, 1985). (Heimlich defines highly erodible land as land that erodes above its tolerance value, even under the best management; this is a condition that is assumed to reflect adoption of the most effective, feasible conservation system.) In contrast, 7.1 percent of all cropland, nearly 30 million acres, meets this definition of highly erodible. When in stable grass or forestry uses, these lands generally erode at very low rates, about 1 to 3 tons/acre·year.

Permanent vegetative cover might be used as a conservation option on particular lands.

Alternative Land Classification Schemes

Enormous pressure exists today to substantially reduce federal farm program expenditures that have cost over $15 billion annually in recent years and are projected to remain over $10 billion annually throughout the next several years. Several options to reduce these program costs through new conservation initiatives have been studied.

One option, the so-called sodbuster provision that was proposed as part of the 1985 Farm Bill, would deny specific USDA program benefits to farmers who cultivate highly erodible land that has not been cultivated for a period of five years. Another option, the conservation reserve, would offer land rental payments to farmers who voluntarily retire erosion-prone land currently in cultivation and put it to long-term, soil-conserving uses.

Findings from analyses of the NRI data illustrate the high levels of erosion concentrated on limited land areas. To make the adoption of such options more administratively feasible and more cost-effective, the committee concludes that an alternative land classification system should be developed that more accurately classifies cropland according to its susceptibility to erosion and erosion-induced damages.

Proposed Options to the LCCS

National and regional analyses of the 1982 NRI demonstrate that the choice of a land classification scheme and criteria are of critical importance in considering new conservation policy initiatives. The committee believes that the 1982 NRI—the USLE data, in particular—constitutes an adequate technical basis for the design and implementation of alternative classification schemes.

A number of new schemes have been proposed in recent years, based on combinations of the factors in the USLE, soil loss tolerances, or the WEE (Heimlich and Bills, 1986). The committee recognizes that these are among a number of possible options that could be formulated. They are noted here because they illustrate the kinds of information needed, the element of judgment involved, and the degree to which the geography and size of areas designated in various erosion classes depends upon the data and equations used.

One approach, originally advanced by the American Farmland Trust, categorizes land into one of three groups based upon progressively higher ranges of inherent potential for sheet and rill erosion (RKLS value) and proposes comparable groupings based on the potential for wind erosion. Another system, developed by Heimlich and Bills (1986), uses RKLS ranges that differ from those in the American Farmland Trust system and different assumptions regarding C- and P-factor values.

Table 5-6 contains a comparison of the total cropland acreage that would be considered highly erodible nationwide under several alternative systems. Depending on the criteria and system applied to the 1982 NRI, analyses showed that between 24 million and 89 million acres of cropland would be considered erosion prone, or highly erodible. The scheme under most intensive study by USDA would include between 32 million and 65 million acres of land considered eligible for a conservation reserve program, depending upon whether the criterion is RKLS/T > 10 or RKLS/T > 15 and whether the actual erosion exceeds twice the relevant soil loss tolerance limit (D. G. Burns, USDA, personal communication, 1985).

The USDA approach improves upon RKLS-based systems only to the extent that T-factor values accurately reflect a soil's susceptibility to erosion. Notwithstanding the known shortcomings of T values, as

TABLE 5-6 Acreage of Highly Erodible Cropland as Calculated Under Alternative Land Classification Criteria

Option	Land Classification Criteria	Acres (million)
	USLE > 2T	50.9
	WEE > 2T	35.4
	LCCS (IVe, VIe, VIIe, VIII)	49.4
AFT[a]	RKLS > 75	24.0
	RKLS/T or CI/T > 10	89.0
	RKLS/T or CI/T > 15	49.7
USDA	RKLS/T or CI/T > 10, eroding > T	64.6
	RKLS/T or CI/T > 10, eroding > 2T	52.7
	RKLS/T or CI/T > 15, eroding > T	36.9
USDA	RKLS/T or CI/T > 15, eroding > 2T	32.0

[a] American Farmland Trust.

SOURCE: Based on 1982 NRI data.

noted earlier, the committee believes that the inclusion of concepts relating potential erosion to productivity is a step in the right direction. However, improved calculations of T values, incorporating the results of the PI and EPIC models, are needed. In particular, the common range of T-factor values (currently 2 to 5 tons/acre·year) should be refined in accordance with the true susceptibility of soils to erosion-induced productivity losses. Improving the accuracy of the relationship will improve such classification schemes.

An advantage of the proposed USDA system is that it can be developed and implemented for most policy purposes by using existing information in soil surveys and the 1982 NRI, supplemented by minimal field work. In addition, improvements in the system—such as those expected from improved estimates of the relation of erosion to potential productivity—can and should be readily incorporated into the system without requiring new policies or altering the impact of new conservation policy initiatives.

The committee believes that further analysis is needed to fully evaluate the suitability of alternative land classification schemes used to identify highly erodible or fragile lands, particularly in relation to policy initiatives. Improper classification of croplands could seriously undermine the effectiveness of a program initiative. For example, confidence limits must be calculated for NRI acreage estimates at the MLRA and smaller levels of aggregation. The geographic distribution of land that would be subject to alternative definitions should be assessed. The implications of a provision that would exempt land cultivated even once during a designated grace period from sodbuster sanctions requires additional study.

Similar concerns arise in consideration of conservation reserve programs. Like the sodbuster provision, to the extent that erosion is of major concern, the eligibility criterion for placing land into a conservation reserve should employ a land classification system based on the inherent potential and actual erosion of the land rather than a land capability class and subclass designation.

Classification schemes based on erosion equations are imperfect. Their basic shortcoming is that gross erosion rates do not always accurately reflect the effects of erosion on soil productivity or the degree of offsite damages from erosion. Ideally, future classification systems for erosion hazards can be based on more explicit criteria, such as PI and EPIC model results. In addition, similar explicit criteria will be needed to classify land, at least in part, according to its potential to cause nonpoint pollution problems.

References

American Farmland Trust. 1984. Soil Conservation in America: What Do We Have to Lose? Washington, D.C.
Baker, D. B. 1984. Fluvial Transport and Processing of Sediments in Large Agricultural River Basins. EPA-600/S3-83-054. Washington, D.C.: U.S. Environmental Protection Agency.
Brown, G. E. 1983. Information management for conservation decisions. J. Soil Water Conserv. 38:451–454.
Christensen, L. A. 1986. Applications of the National Resources Inventory data to inventory, monitor, and appraise offsite erosion damage. In Soil Conservation: Assessing the National Resources Inventory, Vol. 2. Washington, D.C.: National Academy Press.
Clark, E. H., II, J. A. Haverkamp, and W. Chapman. 1985. Eroding Soils: The Off-Farm Impact. Washington, D.C.: The Conservation Foundation.
Colvin, T. S., J. M. Laflen, and D. G. Erbach. 1981. A review of residue reduction by individual tillage implements. Pp. 102–110 in Crop Production with Conservation in the '80s. Proceedings of the American Society of Agricultural Engineers, December 1–2, 1980. Publications No. 7-81. St. Joseph, Mo.: American Society of Agricultural Engineers.
Crosson, P. 1984. New perspectives on soil conservation policy. J. Soil Water Conserv. 39:222–225.
Crosson, P., and S. Brubaker. 1982. Resource and Environmental Effects of U.S. Agriculture. Washington, D.C.: Resources for the Future.
Foster, G. R. 1982. Channel Erosion Within Farm Fields. Preprint 82-007. New York: American Society of Civil Engineers.
Foster, G. R. 1986. Understanding ephemeral gully erosion. In Soil Conservation: Assessing the National Resources Inventory, Vol. 2. Washington, D.C.: National Academy Press.
Gianessi, L. R., and H. M. Peskin. 1981. Analysis of national water pollution control policies. II. Agricultural sediment controls. Water Resources Res. 17:803–821.

Gillette, D. A. 1986. Wind erosion. *In* Soil Conservation: Assessing the National Resources Inventory, Vol. 2. Washington, D.C.: National Academy Press.

Heimlich, R. E. 1985. Soil erosion on new cropland: A sodbusting perspective. J. Soil Water Conserv. 40:322–326.

Heimlich, R. E., and N. L. Bills. 1986. An improved soil erosion classification: Update, comparison, extension. *In* Soil Conservation: Assessing the National Resources Inventory, Vol. 2. Washington, D.C.: National Academy Press.

Iowa Geological Survey. 1984. Hydrogeologic and Water Quality Investigations in the Big Spring Basin, Clayton County, Iowa; 1983 Water-Year. Open File Report 84-4. Iowa City.

Johannsen, C. J. 1986. Potential uses of the NRI in state and local decision making. *In* Soil Conservation: Assessing the National Resources Inventory, Vol. 2. Washington, D.C.: National Academy Press.

Kiniry, L. N., C. L. Scrivner, and M. E. Keener. 1983. A Soil Productivity Index Based Upon Predicted Water Depletion and Root Growth. Research Bulletin 1051. Columbia: College of Agriculture. University of Missouri.

Larson, W. E., F. J. Pierce, and R. H. Dowdy. 1983. The threat of soil erosion to long-term crop production. Science 219:458–465.

Lee, L. K., and J. J. Goebel. 1984. The Use of the Land Capability Class System to Define Erosion Potential on Cropland. Report No. 85-1. Washington, D.C.: Assessment and Planning, Soil Conservation Service, U.S. Department of Agriculture.

McCormack, D. E., and R. E. Heimlich. 1985. Erodible Soils: Definition and Classification. Report No. 85-2. Washington, D.C.: Assessment and Planning, Soil Conservation Service, and Natural Resource Economics Division, Economic Research Service, U.S. Department of Agriculture.

Ogg, C. W. 1986. New cropland in the 1982 NRI: Implications for resource policy. *In* Soil Conservation: Assessing the National Resources Inventory, Vol. 2. Washington, D.C.: National Academy Press.

Pierce, F. J., R. H. Dowdy, W. E. Larson, and W. A. P. Graham. 1984. Soil productivity in the Corn Belt: An assessment of erosion's long-term effects. J. Soil Water Conserv. 39:131–136.

Pierce, F. J., W. E. Larson, and R. H. Dowdy. 1986. Field estimates of C factors: How good are they and how do they affect calculations of erosion? *In* Soil Conservation: Assessing the National Resources Inventory, Vol. 2. Washington, D.C.: National Academy Press.

Pierce, F. J., W. E. Larson, R. H. Dowdy, and W. A. P. Graham. 1983. Productivity of soils: Assessing long-term changes due to erosion. J. Soil Water Conserv. 38:39–44.

Renard, K. G. 1986. A discussion of erosion on range and forest lands: Impacts of land use and management practices. *In* Soil Conservation: Assessing the National Resources Inventory, Vol. 2. Washington, D.C.: National Academy Press.

Richards, B. K., M. F. Walter, and R. E. Muck. 1984. Variation in line transect measurements of crop residue cover. J. Soil Water Conserv. 39:60–61.

Rosenberry, P. E., and B. C. English. 1986. Erosion control practices: Impact of actual versus most effective use. *In* Soil Conservation: Assessing the National Resources Inventory, Vol. 2. Washington, D.C.: National Academy Press.

Runge, C. F., W. E. Larson, and G. Roloff. 1986. A midwestern perspective on targeting conservation programs to protect soil productivity. *In* Soil Conservation: Assessing the National Resources Inventory, Vol. 2. Washington, D.C.: National Academy Press.

Sampson, R. N. 1986. Erosion on range and forest lands: Impacts of land use and management practices. *In* Soil Conservation: Assessing the National Resources Inventory, Vol. 2. Washington, D.C.: National Academy Press.

REFERENCES

Skidmore, E. L., and N. P. Woodruff. 1968. Wind Erosion Forces in the United States and Their Use in Predicting Soil Loss. USDA Agricultural Handbook No. 346. USDA Science and Education Administration. Washington, D.C.: U.S. Government Printing Office.

Society for Range Management. 1985. Announcement. Renewable Resources J. 3:5.

Thorne, C. R. 1984. Prediction of Soil Loss Due to Ephemeral Gullies in Arable Fields. Report CER83-84CRT. Ft. Collins: Colorado State University.

U.S. Congress, House. 1981. Natural resources data bases. Subcommittee on Department Operations, Research, and Foreign Agriculture. 97th Cong., 2d sess., June 2. Serial K.

U.S. Department of Agriculture. 1981. National Summary Evaluation of the Agricultural Conservation Program: Phase I. Agricultural Stabilization and Conservation Service. Washington, D.C.

Walker, D. J., and D. L. Young. 1982. Technical Progress in Yields: No Substitute for Soil Conservation. Current Information Series No. 671. Moscow, Idaho: College of Agriculture, University of Idaho.

Walker, D. J., and D. L. Young. 1986. Assessing soil erosion productivity models. *In* Soil Conservation: Assessing the National Resources Inventory, Vol. 2. Washington, D.C.: National Academy Press.

Williams, J. R., R. R. Allmaras, K. G. Renard, L. Lyles, W. C. Moldenhauer, G. W. Langdale, L. D. Myer, and W. J. Rawls. 1981. Soil erosion effects on soil productivity: A research perspective. J. Soil Water Conserv. 36:82–90.

Wischmeier, W. H., and D. D. Smith. 1978. Predicting Rainfall Erosion Losses: A Guide to Conservation Planning. Agriculture Handbook No. 537, USDA Science and Education Administration. Washington, D.C.: U.S. Government Printing Office.

Appendix

TABLE A Data File Field Definitions and Format in the 1982 NRI

Field	Record Location on Tape	Item and Field Name	Description
1	1–5	FIPS code	State and county Federal Information Processing Standard code[1]
2	6–12	PSU no.	Primary Sampling Unit number[2]
3	13	Point no.	Specific point within the PSU where inventory data are collected—a data record is not included in the NRI file for points that are federal, urban and built-up, transportation facilities, water bodies, or streams[3]
4	14–17	Location code	SCS location code[4]
5	18–21	MLRA	Major Land Resource Area, per USDA Agriculture Handbook No. 296 (December 1981)[5]
6	22–29	Hydrologic unit	Water Resource Council hydrologic unit—eight-digit accounting unit number, per U.S. Geological Survey Circular 878-A (1982)
7	30–35	Expansion factor	Number of acres the sample point represents (in 100s)—taking into account sampling procedure and state's census acres; for use when constructing acreage estimates (for categories in which the point falls)[6]
8	36	Ownership of land	1 = private 2 = municipal 3 = county or parish 4 = state

TABLE A (Continued)

Field	Record Location on Tape	Item and Field Name	Description
			5 = federal (not in file)
			6 = Indian tribal and individual trust lands
			0 = not applicable
9a	37	Land capability class	Soils suitability rating for agriculture, between I and VIII—class I soil has few restrictions that limit its use; class VIII soil has limitations that nearly preclude its use for commercial crop production[7]
9b	38	Land capability subclass	Chief limitation of the soil (except when class I): e = erosion; w = water; s = shallow, droughty, or stony; c = climate
10	39	T factor	Soil loss tolerance factor—indicates acceptable level of annual soil loss, between 1 and 5 tons/acre·year
11	40	Prime farmland	Meets prime farmland criteria?:[8] 1 = yes 2 = no
12	41	Degree of erosion	Degree of erosion:[9] 1 = none or slight 2 = moderate 3 = severe
13	42	Nonarable	Nonarable due to past erosion?:[10] 1 = yes 2 = no
14	43	—	Presently zeroed out
15	44	Saline/alkali	Special management need due to saline and/or alkali soil?:[11] 1 = yes 2 = no
16	45	Type of irrigation	Application of water to soils to assist in production of crops (for 1982, or for at least two of last four years): 0 = no irrigation 1 = gravity irrigation 2 = pressure irrigation 3 = gravity and pressure irrigation
17	46	Water source	Source of water for irrigation: 0 = no irrigation 1 = well 2 = pond, lake, or reservoir 3 = perennial stream 4 = lagoon or other waste water 5 = combination
18	47	Water provision	Irrigation provides 1/2 of the water requirements?: 1 = yes 2 = no

APPENDIX

TABLE A (Continued)

Field	Record Location on Tape	Item and Field Name	Description
19	48	Flood prone	Flood-prone area?:[12] 1 = yes 2 = no
20	49	Cover/use, general	General land cover/use based upon specific land cover/use and cropping history:[13] A = cropland B = pastureland and native pasture C = rangeland D = forestland E = other lands (farms) F = barren lands G = other lands H = urban and built-up land I = rural transportation J = water (census) K = small water bodies (noncensus)
21	50	Cover/use, major	Codes for major category of use[14]
22	51–53	Cover/use, specific	Code by crop[15]
23	54–55	Use	Use of land or water[16]
24	56–58	Cropping history 1	Specific land cover/use for 1981[17]
25	59–61	Cropping history 2	Specific land cover/use for 1980
26	62–64	Cropping history 3	Specific land cover/use for 1979
27	65	Double-cropped	Double-cropping used?:[18] 1 = yes 2 = no
28	66–68	Conservation practice 1	Conservation practices that are currently in use on the land—up to three practices could be identified; whether the practices were identified in fields 1, 2, or 3 is not relevant to priority[19]
29	69–71	Conservation practice 2	See Conservation practice 1
30	72–74	Conservation practice 3	See Conservation practice 1
31	75–76	Conservation need	Conservation treatment needed[20]
32	77–78	K factor	Soil erodibility factor for USLE[21]
33	79–81	R factor	Rainfall factor for USLE
34	82–87	C factor	Cropping management factor for USLE
35	88–90	P factor	Erosion-control practice factor for USLE
36	91–94	Slope length	Slope length (in feet) for USLE
37	95–98	Slope percent	Slope percent for USLE

TABLE A (Continued)

Field	Record Location on Tape	Item and Field Name	Description
38	99	USLE flag	Modified version of USLE applied: 1 = modified LS, for frozen soil 2 = version for R-1 and R-2 areas 0 = none, or not applicable
39	100–103	USLE	USLE calculation of estimated average annual soil loss due to sheet and rill erosion (in tons/acre·year)
40	106–114	USLE tons	Estimated average annual tons of soil loss due to sheet and rill erosion (in 100s of tons/acre·year)
41	115–120	Wind erosion	Estimated average annual soil loss due to wind erosion (in tons/acre·year)
42	121–129	Wind erosion	Estimated average annual tons of soil loss due to wind erosion (in 100s of tons/year)
43	130–131	Dominant soil and water problem	Dominant soil- and water-related problem inhibiting or preventing conversion of land to cropland[22]
44	132–133	Secondary soil and water problem	Secondary soil- and water-related problem inhibiting or preventing conversion of land to cropland[23]
45	134–135	Dominant other problem	Dominant other problem inhibiting or preventing conversion of land to cropland[24]
46	136–137	Type of effort necessary	Effort necessary for conversion to cropland:[25] 01 = none: can convert by beginning tillage 02 = on-farm: can convert through actions by individual farmers 03 = multifarm: informal or formal cooperation between neighbors to install systems 04 = project action required 09 = not applicable, e.g., zero potential 99 = currently cropland, built-up, transportation, or water or VIIe, VIIw, VIIs, or class VIII land
47	138–139	Conversion potential	Potential for conversion to cropland within the foreseeable future:[26] 00 = zero potential 01 = conversion unlikely in foreseeable future 02 = medium potential 03 = high potential 99 = currently cropland, built-up, transportation, or water or VIIe, VIIw, VIIs, or class VIII soil
48	140–141	Type of wetland	Wetland type: 00 = not a wetland type 1–20 01–20 = type, as per circular 39, Department of Interior

TABLE A (Continued)

Field	Record Location on Tape	Item and Field Name	Description
49	142	Kind of vegetation	Kind of wetland vegetation: 0 = none 1 = emergent 2 = scrub/shrub 3 = forested
50	143	Kind of wetland system	A complex of wetland and deep water habitats influenced by hydrologic, geomorphological, chemical, and/or biological factors: 0 = no kind 1 = marine 2 = estuarine 3 = riverine 4 = lacustrine 5 = palustrine
51	144	Riparian area kind	Kind of riparian area:[27] 0 = none 1 = natural streambank 2 = man-made canal or ditch bank 3 = natural pond or lake shoreline 4 = man-made pond or reservoir 5 = tidal area shoreline
52	145	Riparian vegetation	Kind of riparian vegetation: 0 = none (nonriparian area or barren) 1 = trees 2 = shrubs 3 = forbs 4 = grass and grasslike plants 5 = mixed 6 = other
53	146	Riparian width	Width of the strip of riparian vegetation: 0 = none 1 = less than 100 feet 2 = 100–500 feet 3 = greater than 500 feet
54	147–150	Distance to cropland	Distance from point to nearest occurrence of cropland (in feet); 9999 if nearest occurrence is further than 5,280 feet[28]
55	151–154	Distance to forestland	Distance from point to nearest occurrence of forestland (in feet)
56	155–158	Distance to grassland	Distance from point to nearest occurrence of pastureland or rangeland (in feet)
57	159–162	Distance to water	Distance from point to nearest occurrence of water (in feet)
58	163–166	Distance to wetland	Distance from point to nearest occurrence of wetlands type 1–20 (in feet)

TABLE A (Continued)

Field	Record Location on Tape	Item and Field Name	Description
59	167–170	Distance to built-up	Distance from point to nearest occurrence of farmsteads, urban and built-up, roads, etc. (in feet)
60	171	Winter cover kind	Winter ground cover of the last harvested crop: 0 = none or not cropland 1 = live crop 2 = cropland residue
61	172–173	Winter cover height	Height of crop or residue remaining over winter (in inches)
62	174	Upright residue	Residue remains upright over winter?: 1 = yes 2 = no 0 = not cropland
63	175	Pastureland condition	Pastureland condition rating:[29] 2 = good 3 = fair 4 = poor 9 = not applicable 0 = not pastureland
64	176	Woody canopy cover, for pastureland	Canopy cover of woody plants, if pastureland or native pasture: 1 = 0–10% 2 = 10–25% 3 = 26–55% 4 = 56–100% 0 = not pastureland
65	177	Rangeland condition	Rangeland condition rating (percent climax vegetation): 1 = excellent (76–100%) 2 = good (51–75%) 3 = fair (26–50%) 4 = poor (0–25%) 8 = annual range 9 = not applicable 0 = not rangeland
66	178	Woody canopy cover, for rangeland	Canopy cover of woody plants, if rangeland: 1 = 0–9% 2 = 10–25% 3 = 26–55% 4 = 56–100% 0 = not rangeland
67	179	Rangeland condition trend	Apparent trend in condition of the soil and/or vegetation resources, for rangeland: 1 = up (soil and/or vegetation resources improving)

APPENDIX

TABLE A (Continued)

Field	Record Location on Tape	Item and Field Name	Description
			2 = even (not readily apparent)
			3 = down (resources deteriorating)
			0 = not rangeland
68	180	Grazing level, for rangeland	Status of grazing, for rangeland:
			1 = not routinely grazed
			2 = routinely grazed but presently deferred
			3 = currently grazed, lightly to properly used
			4 = currently grazed, excessively used
			0 = not rangeland
69	181	Forest type, general	General category of forest cover type (blank if not forest)[30]
70	182–184	Forest type, specific	Specific category of forest cover type
71	185	Canopy cover, for forestland	Canopy cover of trees, for forestland:
			1 = 0–9%
			2 = 10–25%
			3 = 26–55%
			4 = 56–100%
			0 = not forestland
72	186–188	Basal area/stem count	Basal area of the stand (in square feet per acre), if average diameter at breast height (DBH) is at least 5 inches; if less than 5 inches, then stocking based on stem count is given:
			900 = poorly stocked
			901 = moderately stocked
			902 = fully stocked
			000 = nonstocked or not forestland
73	189–191	DBH	Average DBH, for forestland (in inches)
74	192	Forest understory composition	Primary plant group for forestland understory composition:
			0 = none
			1 = woody
			2 = forbs
			3 = grass and grasslike plants
75	193	Understory forage value	Forage value rating of understory for forestland, based upon percentage of understory production by preferred species:
			1 = very high (51–100%)
			2 = high (31–50%)
			3 = moderate (11–30%)
			4 = low (0–10%)
			9 = not applicable, i.e., not suitable for grazing
			0 = not forestland
76	—	Soils-5	Soils-5 identification block; not standard field on NRI data file[31]

SOURCE: 1982 NRI.

Notes to Appendix

[1] The Federal Information Processing Standard (FIPS) code identifies the state and county in which the sample point falls.

[2] The Primary Sample Units (PSUs) compiled by the Iowa State University Statistical Laboratory were located by field personnel on aerial photographs. The PSUs were then examined to determine if the inventory could proceed based on existing soil surveys or if old surveys should be updated or new surveys made. Boundaries of the PSUs were transferred to the most recent aerial photograph on which soil characteristics had been mapped.

[3] Most PSUs have three sample points, although smaller (40-acre) PSUs have only one. The locations of these points were given on a gummed label sent from the Iowa State University Statistical Laboratory. Two coordinates were given to locate each point.

[4] Internal SCS code for use with time and progress reporting system.

[5] The United States is divided into 156 Major Land Resource Areas (MLRAs), some of which are further divided into subareas denoted by letters (i.e., MLRA 83A is the northern Rio Grande Plain, 83B is the western Rio Grande Plain, and so forth).

[6] Expansion factors were assigned by the Iowa State University Statistical Laboratory. This factor specifies the number of acres that the sample point represents for constructing acreage estimates for categories in which the point falls. Fields 8 to 39, 41, and 43 to 76 describe characteristics of the point. When estimated erosion rates are expanded to geographic aggregates larger than the land represented by the sample point, use fields 40 and 42.

[7] The Land Capability Class System (LCCS) is described in Agriculture Handbook No. 210, SCS, USDA.

[8] Prime farmland is land that has the best combination of physical and chemical characteristics for producing food, feed, forage, fiber, and oilseed crops and is also available for these uses (the current use could be cropland, pastureland, rangeland, forestland, or other land, but not urban built-up land or water). Specific criteria for prime farmland are given in the Federal Register, Vol. 43, No. 21, January 31, 1978.

[9] Degree of erosion: (1) None or slight—Accelerated erosion has not greatly altered the thickness and character of the A horizon. There may be a few rills, some deposits of windblown sediment near plants or clods or places with thin A horizons that indicate slight accelerated erosion is taking place. (2) Moderate—Accelerated erosion has reduced the thickness and character of the A horizon. In cultivated areas, erosion has removed enough of the original A horizon so that tillage or other implements have mixed the original A horizon and underlying horizons. In uncultivated areas, approximately 25 to 75 percent of the original surface soil has been removed by erosion from most of the area. There may be a few shallow gullies, scoured or blown-out areas, or evidence of soil drifting. (3) Severe—The soil has been eroded to the extent that all or practically all of the original surface soil has been removed. The surface layer consists essentially of materials from the B horizon or other underlying horizons. Severe gullying, scouring, drifting, or dune development is included.

[10] Yes is answered if past erosion had changed land formerly suitable for cropland to marginally suitable or unsuitable for cropland use. This question was answered for all land uses and types of vegetative cover.

[11] A saline soil interferes with the growth of most crop plants, because of high salt content. An alkali soil reduces the growth of most crop plants because of high sodium or high salt and sodium. Such soils require special management practices and measures for reclamation.

[12] Flood-prone areas adjoin rivers, streams, watercourses, bays, lakes, alluvial fans and plains, or other areas that in the past have been covered intermittently by floodwater or could be expected to be flooded in the future. Upland depressions are not included.

APPENDIX

Flood-prone areas are the approximate areas subject to inundation by a flood having an average recurrence interval of once in 100 years (floods having a 1 percent chance of occurring in any given year). The area was to be determined on the basis of sound engineering analyses where available and on interpretative analyses where engineering analyses are not available. Sources of information included studies by the SCS, the Corps of Engineers, and the U.S. Department of Housing and Urban Development and maps prepared by the U.S. Geological Survey, the Tennessee Valley Authority, and the National Oceanic and Atmospheric Agency.

[13]General land cover/use in 1982 is coded in field 20 as follows: A, cropland/hayland; B, pastureland, native pasture; C, rangeland and tundra; D, forestland; E, other land in farms; F, barren land; G, other lands; H, urban and built-up, small built-up; I, rural transportation; J, water, census; K, water body, smaller than 40 acres. In cases where the point fell on a land cover/use boundary, field personnel first tried to identify the land cover/use to the north of the point. If this also was a boundary, the land cover/use to the east of the point was recorded. The same rule was applied if the point fell on a fence row or a narrow waterway.

[14]Field 21 contains the next level of land cover/use classification, an alphabetic code for major land cover/use. The letter A will appear in this field for all pastureland and native pasture, rangeland and tundra, forestland, other land in farms, barren land, other lands, rural transportation, and census-sized water areas. For cropland, urban and built-up land, and small water bodies, however, field 21 will contain one of the following alphabetic codes for major land cover/use. For cropland: A, horticultural crops; B, row crops; C, close-grown crops; D, hayland; E, other. For urban and built-up: A, urban/built-up (less than 10 acres); B, small built-up (0.25-10 acres). For water body: A, water body smaller than 40 acres; B, small perennial stream.

[15]This and other codes can be found in the 1982 NRI field documentation guide (SCS, USDA).

[16]The numeric code entered in this field reflects the primary or dominant use of land, regardless of land cover/use recorded above. For example, in a rural park, the land cover/use item may be forests, not grazed (coded 342 in field 22), but the use of the land would be recorded as recreation (43) in field 23.

[17]Fields 24, 25, and 26 specify land cover/uses for 1981, 1980, and 1979, respectively. The same codes are used here as in field 22 (see note 15). For example, if the land was in continuous rangeland cover/use, the numeric code 250 would appear in each of these fields. A 1979-1981 cropping history of corn, soybeans, and corn would be coded in these fields as 011, 013, and 011, respectively. Information on crop rotations was used in the Wind Erosion Equation and Universal Soil Loss Equation (USLE). Including the entry for the current (1982) crop, four years of cropping history are available altogether.

[18]This field applies to cropland that was used for horticultural, row, or close-grown crops. No was entered for all other land cover/use codes, including water. If information for the current year (1982) was limited, the entry reflects a history of double-cropping in the field for two or more years out of the last four years. Hay and pasture was not considered double-cropped.

[19]Fields 28 to 30 record up to three conservation practices in use in 1982. The practices must have been applied to the area in which the point fell or in the portion of the field surrounding the point that would be used for conservation planning. The point need not have fallen on the specific practice to be recorded. For example, if a point fell on a farmstead that had a farmstead and feedlot windbreak, code 380 should appear in field 28. If none of the listed practices were encountered, 000 should be entered for each field. If only one practice was in use, it is coded in field 28 and the other fields should show 000. Whenever terraces are coded (600) for one field, it is presumed that contour farming was in use, so code 330 should appear in another of the three fields.

[20]The data were obtained for the area in which the point fell or for the portion of the field surrounding the point that would be used in conservation planning. Conservation needs were based on the judgment of a qualified specialist or technician as guided by the local technical guide, prevailing agricultural operations, and the practical bases and guides used and exercised in the development of conservation plans. Primary or dominant treatment needs were recorded. Field personnel were instructed to give erosion control treatment needs priority over other needs such as treatment to increase wood or forage production.

[21]Fields 32 to 37 contain values for the six factors in the USLE. Instructions referred field personnel to local technical guides, SCS Technical Service Center publications, and Agriculture Handbook No. 537 (December 1978) for detailed explanation of factors in the USLE. Determinations were made for the field in which the point fell or the portion of the field surrounding the point that would be used in conservation planning.

Field 32: K factor. The erodibility factor for the soil series. Specialists were to estimate K factors for mapping units that were complexes, associations, undifferentiated units, or miscellaneous land types or for mapping units above the soil series level.

Field 33: R factor. The current rainfall factor map was used. Where appropriate—mainly in the Pacific Northwest—a thaw-snowmelt adjustment was included. This adjustment is encoded in field 38.

Field 34: C factor. The cropping management factor reflects crop sequence and residue for cropland. Areas in pastureland, rangeland, and forestland cover/uses reflect a different extension of the factor than cropland. Percent ground cover was determined from a perpendicular view.

Field 35: P factor. The value for the erosion control practice factor was entered from Tables 13, 14, and 15 of Agriculture Handbook No. 537. If the value in this field is less than 1.0, the value and then a code of either 330 (contour farming) or 585 (contour stripcropping) should be entered in fields 28 to 30.

Field 36: Slope length. The length of slope (in feet) through the point. On terraced land, the distance between terraces was entered. Slope length is the distance from the point of origin (whether on or off the PSU) of overland flow to either of the following: (1) the point where the slope decreases to the extent that deposition of sediment begins or (2) the point where runoff enters an area of concentrated flow or a channel.

Field 37: Percent slope. Rounded to the nearest percent on slopes greater than 1 percent and to the nearest 0.1 percent on slopes of less than 1 percent. Zero should not appear. Slope percent was measured on the segment of the land form on which the point fell and in the direction that water would flow overland.

[22]Fields 43 to 47 are coded 99 if the land at a sample point was urban and built-up, rural transportation land, water, or had a land cover/use of cropland; capability class and subclass VIIe, VIIw, or VIIs; or capability class VIII. If the point is not coded 99, personnel determined the dominant reason that would inhibit or prevent conversion of the land to cropland.

[23]The same code as used in field 25 is used here to record secondary soil and water problems that would inhibit or prevent conversion of land to cropland.

[24]Other reasons that would inhibit or prevent conversion of land to cropland were of a social, economic, or environmental nature.

[25]The type of effort necessary for conversion to cropland was determined to provide a relative measure of the magnitude and time that might be involved in converting land to cropland.

[26]Upon completion of the determinations encoded in fields 25 to 28, SCS personnel met with representatives from other agricultural agencies in the county, including the Agricultural Stabilization and Conservation Service, Economic Research Service, Farmers Home Administration, and the Forest Service. The group determined for each point the

APPENDIX

potential for conversion to cropland within the forseeable future (10 to 15 years). Determinations were based on commodity prices, development costs, and production costs for the year prior to the current year. High potential was to be assigned in cases where similar land had been converted to cropland during the preceding three years. Lands held for urban or related development were excluded from medium and high potential.

[27]A riparian area was defined as the bank, shoreline, or edge of the rising ground bordering a natural or constructed watercourse or a water area (lake or tidal area, for example). Riparian was not limited to natural areas.

[28]Fields 54 to 62 contain data with which to relate wildlife habitat to land cover, use of an area, and diversity of an area in which the sample point fell. For fields 54 to 59, 0 was entered if the point fell in the respective diversity types. Diversity was not shown beyond 5,280 feet (1 mile) from the sample point. If a given diversity type either did not exist or was beyond 5,280 feet, the field should be coded 9999.

[29]Definition of codes for rating pastureland and native pasture are as follows:

2—Good. Best suited plants are being used. There is moderate- to high-level fertility and good to excellent grazing management. Grazing is at an intensity for maximum plant production and vigor.

3—Fair. A moderate level of management is being used. Plants are adapted to climate and soils but not necessarily for the best designated use. Grazing is at an intensity that limits production to moderate levels. Erosion is minimal. Fertilization is irregular and unplanned. A continuous grazing system is in use.

4—Poor. There is improper pasture use or a low level of management, or plants are not well suited to climate and soil. The soil has a low fertility level and evidence of erosion. Brush management may be necessary.

9—Not applicable. Production is by native species not routinely fertilized, overseeded, or irrigated.

[30]The point was considered forestland if trees were recently harvested but not currently stocked or committed to some other use.

[31]In Chapter 2 the committee recommends that a supplementary tape be issued containing the Soils-5 identification code and that future inventories contain these data.

Index

A

Aggregate-level research, 68
Agricultural Research Service, 25, 41, 68, 72
Agricultural Runoff Model, 53
Agricultural Stabilization and Conservation Service, 41
Agricultural Department (USDA), xv, 19, 30. *See also* Soil Conservation Service

B

Bureau of Land Management, 30

C

C (correction) factor, WEE, 54, 57
C (vegetative cover and management) factor, USLE, xviii, 37–38, 40–52, 78
Chemicals, Runoff, and Erosion from Agricultural Management Systems, 53
Clean tillage, 16
Concentrated flow erosion, 34. *See also* Ephemeral gully erosion
Conservation Needs Inventory, 4
Conservation practices, xvi
  contour farming, 16, 38, 53
  conservation tillage, 16, 47–48, 60, 76, 78–81, 84
  distribution and effectiveness, 16–18, 53, 75–76
  ephemeral gully erosion, 60
  erosion reduction goals, 83–84
  grassland waterways, 36, 53, 59, 60
  minimum tillage, 16
  no-till farming, 16, 50, 53
  NRI applications, 3, 82
  P factor, 36, 38
  permanent vegetative cover, 90–91
  stripcropping, 16, 36, 38, 53
  targeting of activities, 83
  terracing, 16, 36, 38, 53, 60
  water contamination problems, 32
Conservation reserve program, 6, 91, 92, 94
Conservation tillage
  definition, 16
  distribution and effectiveness, 16, 47–48, 76, 78–81
  ephemeral gully erosion, 60
  water quality concerns, 84
Contour farming, 16, 38, 53
Cooperative State Research Service, 68
Corn Belt, 11, 12, 76, 77, 81
Corn production, C factor, 45
Cropland, 8
  C factor, 40, 45
  class IIIe, variations in erosion potential, 86–88
  concentration of erosion, 13–16, 81–82
  erosion rates, 8, 11, 12
  land converted to, erosion and productivity problems, 90–91
  T values, 9

D

Data compilation and dissemination
  additional data requirements, xix
  distribution of NRI computer tapes, 25–26
  presentation of published data, 23–25
  remote sensing information, xix–xx, 26–30
  statistical documentation needs, 21–23
  supplemental NRI data, xvii

upgrading SCS inventory and monitoring functions, 21
water quality data, 3-4, 31-33
Defense Department, 30
Drinking water, xix

E

Economic Research Service, 41, 68, 72
Energy Department, 30
Environmental Protection Agency, 4, 30
Ephemeral gully erosion
 control methods, 60
 definition, 34-35, 59
 estimation methods, 59-60
 extent and magnitude of, 60
 and land classification discrepancies, 88
 NRI applications, xvi-xvii, xix
 research needs, 60-61
Erosion effects, 62
 long-term production costs, 64-66
 long-term productivity, 67-68
 short-term production costs, 63-64
 social costs, 72
 technological change and, 66-67
 water quality, 72-74
Erosion Productivity Impact Calculator (EPIC), 53, 69, 72, 83, 93
Erosion-productivity models, xviii, 52-53, 68-69, 74, 84
 EPIC, 53, 69, 72, 83, 93
 PI, 53, 69-71, 83, 93
Erosion-productivity relationship, xviii, xix, 3
Erosion Reconnaissance Survey of 1936, 1
Erosion studies, 34
 erosion prediction equations, xvi-xvii
 erosion trends, 1979-1982, 2
 highly erodible land, definition, 91
 historical perspective, xiii-xiv
 NRI applications, xvi
 soil formation and loss rates, 8
 See also National Resources Inventory (NRI), 1982
Extension Service, 68

F

Farm program expenditures, 91
Federal lands, NRI extension to, xv, xix, 30-31

Forest Service, 30, 41
Forestland, 4, 8
 C factor, 45
 erosion rates, 8, 11
Fragile Soils Work Group, USDA, 59, 84

G

Georgia, 76, 77
Grassed waterways, 36, 53, 59, 60

H

Hay cropland, 38, 40, 91
Hydrogeological data base, xvii
Hydrologic Simulation Program in Fortran, 53

I

I (soil erodibility) factor, WEE, 54, 57
Inherent erosion potential, 40-41
Interior Department, 30
Inventory function, SCS, 19, 20
Iowa, 11, 12, 76, 77
Iowa Geological Survey, 32
Iowa State University, 23

K

K (soil erodibility) factor, USLE, 36-37
K (soil ridge roughness) value, WEE, 55

L

L (length of slope) factor, USLE, 36, 37
L (unsheltered distance) factor, WEE, 55
Land Capability Class System (LCCS), 43, 85-94
 class IIIe, variation in erosion potential, 86-88
 description, 23, 85
 improvement recommendations, 86
 limitations and inconsistencies, 85-86
 proposed alternatives to, 91-94
 regional analyses of inconsistencies, 88-90
 usefulness in conservation planning, 90
Land classification schemes, xvi
 according to erosion rates, 3, 23, 41, 92
 according to wind erosion hazard, 59, 92
 proposed systems, 92-94
 shortcomings, 94

INDEX

M

Major Land Resource Areas (MLRAs), 41
  MLRA 103, 42, 50, 70, 71, 79, 81
  MLRA 105, 42, 50, 71, 79
  MLRA 108, 70, 71
  MLRA 109, 71
  MLRA 113, 71
  MLRA 115, 70, 71
  MLRA 134, 42, 79, 81
  MLRA 136, 42, 50, 79, 88, 89
Megarill erosion, 34. *See also* Ephemeral gully erosion
Minimum tillage, 16
Minneapolis, 57
Missouri, 26, 83
Models. *See* Erosion-productivity models
Monitoring function, SCS, 19, 20
Mulch-factor value, 45-47

N

National Cooperative Soil Survey Program, 85
National Oceanic and Atmospheric Administration, 4, 30
National Resources Inventories (NRIs), 19-33
  cost considerations, 20
  distribution of computer tapes, 25-26
  extension to federal lands, xv, xix, 30-31
  improvement recommendations, overview of, xiv-xx, 19-20, 78
  presentation of published data, xvii, 23-25
  sensing technologies use in, xix-xx, 26-30
  statistical documentation needs, 21-23
  upgrading SCS inventory and monitoring functions, 21
  water quality issues, 31-33
National Resources Inventory (NRI), 1977, 1, 83
  data publication, xvii
  1982 NRI data compared, 4-5
  sheet and rill erosion data, 81
  Soils-5 file linkage, 26
  wind erosion estimates, 54, 56, 88
National Resources Inventory (NRI), 1982
  concentration of erosion, 13-16
  content and methodology, xiv, 1-2, 19, 98-108
  cost, 1
  data publication, xvii, 24
  identification of needs and opportunities, 3
  land use data, 7-8
  LCCS inconsistencies, 85-94
  1977 NRI data compared, 4-5
  soil classification according to erosion rates, 3
  soil conservation practices on cropland, 16-18, 75-76
  soil erosion rates, 8
  Soils-5 file linkage, 26
  T values, 8-12
  as tool for research and policy analysis, 6-7, 82
  trend analysis, 2
  water quality issues, 3-4, 73-74
  wind erosion estimates, xvi, 54-56
National Technical Centers, SCS, 23
Nitrate contamination, xix, 4
Nonpoint Source Model, 53
No-till farming, 16, 50, 53

O

Ohio, 77

P

P (supporting conservation practices) factor, USLE, 36, 38, 40
Pastureland, 11, 91. *See also* Rangeland
Policy options, xvi, 4, 6-7, 82
  conservation reserve, 6, 91, 92, 94
  erosion reduction goals, 83-84
  sodbuster provision, 6, 91
  targeting of conservation activities, 83
Production costs. *See* Erosion effects
Productivity index (PI) model, 53, 69-71, 83, 93

R

R (rainfall) factor, USLE, 36, 88
Rangeland, 8
  C factor, 48
  erosion rates, 8, 11, 12
  federal land, NRI extension to, 31
  T values, 9, 11
  USLE unsuitability to, 36
Reduced tillage, 16

Remote sensing, xix–xx, 26–30
Resource Conservation Act (RCA), 72
  C factor, 40, 45, 50
  concentration of erosion, 13–16, 81–82
  effects on erosion, xviii
  erosion rates, 8
Runoff, xix, 32, 36

S

S (steepness of slope) factor, USLE, 36, 37
Science and Technology Policy, Office of, 30
Sheet and rill erosion
  class IIIe cropland, 86–88
  concentration of, on cropland, 13, 14, 81–82
  inherent erosion potential, 40–41
  NRI applications, 3
  rates, 8, 11, 12
  T value relationship, 11
  See also Universal Soil Loss Equation
Sodbuster policies, 6, 91
Soil and Water Resources Conservation Act of 1977, 1, 19–20
Soil Conservation Act of 1935, xiii
Soil conservation reserve, 6, 91, 92, 94
Soil Conservation Service (SCS), xiii, 1
  distribution of computer tapes, 25–26
  EPIC model, 72
  erosion-productivity research, 68
  guidelines on inherent erosion potential, 41
  presentation of published data, 24–25
  statistical documentation needs, 21–23
  upgrading of inventory and monitoring functions, 19, 20
  USLE improvements, 53
Soil loss tolerance limits (T), xviii, 8–12
Soils-5 file, xvii, 3, 69
  NRI linkage, 25–26
Statistical documentation needs, 21–23
Stripcropping, 16, 36, 38, 53
Surface runoff, xix, 32, 36

T

T (tolerance) values, xviii, 8–12, 84, 93
Technological change
  erosion effects and, 66–67
Terracing, 16, 36, 38, 53, 60

U

Universal Soil Loss Equation (USLE), xvi, xvii, 19, 24, 34–53
  C-factor values, uncertainties associated with, 44–48
  C factor, role in controlling erosion, 41–44
  ephemeral gully erosion, 59
  improvement recommendations, 48, 52–53
  mulch-factor value, 45–47
  1982 NRI values, 38–41
  sheet and rill erosion applications, 34, 35
  simulated erosion rates, 48–52
  variables defined, 36–38
  water pollution problems, 74
U.S. Geological Survey, 4, 30, 32

V

V (vegetative cover) factor, WEE, 55
Vegetative cover
  C factor, USLE, xviii, 37–38, 40–52, 78
  as conservation practice, 90–91
  grassed waterways, 36, 53, 59, 60

W

Water pollution
  erosion damage, 72–73
  NRI applications, xviii–xix, 3–4, 25, 31–33, 73–74
Wind erosion, 81
  concentration of, on cropland, 13, 15, 16
  and land classification discrepancies, 88
  NRI estimates, xvi
  rates, 8
  T value relationship, 11
Wind Erosion Equation (WEE), xvi, xvii, xix, 19, 24, 26, 34, 54–59
  improvement recommendations, 56–59
  limitations, 56
  variables defined, 54–55

Y

Yield Soil Loss Simulator, 72